ALFRESCO FLIGHT

The RAAF Antarctic Experience

DAVID WILSON

Royal Australian Air Force Museum

An occasional series Number 3

Other titles in this series:

Secret Action of 305
Smith & Coghlan
Winner of the 1988 Heritage Award

The RAAF Mirage Story
Compiled by Wing Commander M.R. Susans
Winner of the 1989 Heritage Award

Alfresco Flight
The RAAF Antarctic Experience
David Wilson
Royal Australian Air Force Museum
An Occasional Series
Number 3

First published in 1991 by
Royal Australian Air Force Museum
RAAF Base
Point Cook, Vic., 3029, Australia

Cover and text design by David Constable
Production by Sylvana Scannapiego, Island Graphics
Typeset in Baskerville by Optima Typesetters
Printed in Australia by Impact Printing

Cataloguing-in-publication information

Wilson, David, 1944–
Alfresco flight.

ISBN 0 642 10522 7.

1. Australia Royal Australian Air Force — History.
2. Aeronautics — Antarctic regions — History. 3.
Antarctic regions — Discovery and exploration. I. Royal Australian Air Force Museum. I. Title.

629.1309989

CONTENTS

Metric conversion

1 inch	=	2.5 centimetres
1 foot	=	0.3 metres
1 yard	=	0.9 metres
1 mile	=	1.6 kilometres

INTRODUCTION

AUSTRALIA and her inhabitants have long historical links with the Antarctic continent and the sub-Antarctic islands, dating back to the days of Captain James Cook. The first grounding on the Antarctic continent was made by a Norwegian resident of Melbourne, H.J. Bull, who landed a party of men from the whaling ship *Antarctic* during January 1895 at Cape Adare. A member of his crew, C.E. Borchgrevink, led the

The R.E.P. Monoplane, minus wings, was used as an 'air sled' by the Mawson Expedition of 1911. (Author's collection)

British Antarctic Expedition of 1898–1900, which was the first party to spend an Antarctic winter on the continent. The first Australian to set foot on the continent, Louis Charles Bernacchi, accompanied this party, and Australians were prominent members of many of the famous exploring parties. Bernacchi accompanied Robert Falcon Scott during his 1901–1903 expedition; Douglas Mawson and Edgeworth David were members of Ernest Shackleton's 1906–1907 expedition. Frank Hurley, the photographer, was a member of Shackleton's ill-fated 1914 expedition, and the photographs he exposed are among the classics of Antarctic photography. It was little wonder that men of the calibre of Mawson would wish to lead a national expedition, and this came to fruition during 1911.

Aviation has also been linked with Antarctic exploration. Scott used a hot air balloon for observation purposes. But the vision of the men and the technical development of aircraft did not coincide for some two decades. It is significant that when they did, the RAAF was in the forefront of the deployment of an aerial presence over the southern continent.

The writing of this account of the RAAF presence in the Australian Antarctic Territory would not have been possible without the assistance of the following: Mike Austin, Air Commodore Norman Ashworth (Rtd), Air Commodore Graham Dyke (Rtd), Evelyn Barrett, Ivan Grove, Peter Clemence, Alan Richardson, Garry Cooper, Stuart Campbell, Sylvia Felton, Bob Piper, David 'Jock' Ross, Rod Reeves, Gerry Sebastian, Ian McGarry, David Vincent, Wayne Brown, Bert Cookson, Mel Davis and Neil Mulligan.

My special gratitude to Chris Coulthard-Clark for his professional guidance and assistance.

CHAPTER ONE

Airmen and explorers

SIR Douglas Mawson is a legend among Antarctic explorers. He was also a man of vision, and was well aware of the possibilities of aviation in the Antarctic – so much so that he ordered an R.E.P monoplane from the Vickers Brooklands works for use during his 1911 expedition to the southern continent. The aircraft, which was designed by Robert Esnault Pelteric, was the second built at Brooklands. In his book *Home of the Blizzard* Mawson states that it was intended 'that so far as the flier was concerned, it would be chiefly exercised for the purpose of drawing public attention to the expedition in Australia, where aviation was then almost unknown.' (1) He must have had doubts as to the efficacy of the frail machine under the weather conditions which would prevail on the Antarctic continent. But, in the event, the aircraft was not destined to be used in the flying role during the expedition.

On 5 October 1911 the pilot, Lieutenant H.E. Watkins, and his passenger, Frank Wild, took off from the Cheltenham Raceway at Adelaide on a short flight. Short it was, as the aircraft only flew for three to four minutes to a height of 150 feet, before dropping its left wing and crashing. Fortunately no one was injured, but the aircraft could not be utilized in its prime role and was shipped to Hobart in a wingless state. From Hobart it was transported to the Antarctic with the expedition and, maintained by F.H. Bickerton, was used on several occasions as an air sled. It was finally abandoned some ten miles from winter quarters at Cape Denison.

Mawson did not discount the possibility that aircraft could prove useful for his exploratory and scientific endeavours. The organization of the British, Australian and New Zealand Antarctic Research Expedition (BANZARE) gave him the opportunity to prove new techniques in

The first Australian aircraft to serve over the Antarctic Continent, Gipsy Moth VH–ULD, during the BANZARE 1929–1931. (Author's collection)

Antarctic exploration. Among the equipment was 'a light aeroplane for traversing pack-ice and for increasing the range of observation'. (2) He would have preferred two aircraft, but space aboard the Royal Research Ship *Discovery* precluded carrying more than one. The aircraft selected was a two seat De Havilland DH60G Moth, registered VH-ULD, which was to be the first Australian aircraft to see service over the Antarctic.

Flying Officer S.A.C. Campbell and Pilot Officer G.E. Douglas were seconded from the RAAF for duty with the expedition. Campbell had just completed a short tour aboard the seaplane tender H.M.A.S. *Albatross* and thus had floatplane experience. It is not unusual that service members should be so employed at this period. The RAAF had assisted in a survey of central Australia, and this fact would have been known to Mawson as one of his colleagues had been closely involved with the operation. Both pilots ultimately joined other expedition members aboard the *Nestor* at Melbourne before voyaging to Cape Town to join the *Discovery*.

The *Discovery* was Scott's old ship, and had been supplied by the British government for the expedition. It was stocked for the venture at Cape Town, before departing for Kerguelen Island on 19 October 1929. The voyage gave expedition members an opportunity to gain experience at their various tasks. Campbell, before being bedridden with pneumonia, assisted others in the checking of stores and the restacking of coal briquettes used to fuel the ship. Douglas maintained the motor boat engine, a task which

S.A.C. Campbell, pioneer Antarctic airman, who became the Executive Officer of ANARE 1947–1948. (Author's collection)

G.E. Douglas, Campbell's co-pilot and the leader of the RAAF contingent who discovered the American explorer Lincoln Ellsworth during January 1936. (Author's collection)

was to become one of his extra curricular duties for the duration of the voyage, as well as the Moth. The aircraft was stowed on a grating above the deck, from where two lifeboats had been removed.

The expedition arrived at Possession Island on 2 November to make scientific observations. Sealers were active on the island, and Mawson's men were sickened by the wholesale slaughter of sea elephants. For this reason alone, the departure for Kerguelen on 9 November was welcome.

Kerguelen impressed with its natural, rugged beauty and was the site for many scientific observations. Both Campbell and Douglas participated in these, as well as the more tedious task of re-coaling the *Discovery*. While at Kerguelen, the remaining lifeboat was put ashore to 'clear skids for (the) aeroplane when (it was) assembled in Antarctica'. (3) These duties did not prevent Campbell from accompanying Mawson and seven others on a field trip to Bras d'Bossiere, and the scientific staff found their stay most rewarding. But the main task lay ahead, the *Discovery* departing for Heard Island on 24 November.

Douglas joined the party which went ashore to carry out observations. Campbell, to his disappointment, remained on board to overhaul the echo sounder. Weather conditions were atrocious, with the shore party seeking shelter in an old hut from which they evicted some of the local sea lions. The return to the ship was not uneventful. When on passage to pick up the last shore party, Mawson and Douglas found themselves in jeopardy when they encountered heavy seas in the launch. Visibility was limited by flying spray and fog, and the two men were forced well out to sea to avoid the waves crashing over an offshore reef before heading in the direction of Rogers Head. Some 100 yards from the basalt cliffs, the launch motor cut. A dredging line had to be attached to the anchor rope to prevent the launch from being swept onto the cliffs. Douglas repaired a fuel blockage which had caused the engine malfunction by disconnecting the fuel tank connections. The pitching boat did not make this an easy operation. The connecting nut was dropped under the floorboards of the launch, and anxious minutes were spent in recovering it. Once the repairs were completed, the party ashore was shipped back to the *Discovery* which sailed from Heard Island on 4 December.

Due to inclement weather it was not until 19 December that Campbell and Douglas were able to commence the erection of the Moth. Prior to work commencing, 'early on the morning of December 6, scientists and ships officers gathered outside the cabin of Eric Douglas and serenaded him on the occasion of his birthday'. (4) Despite the efforts of the aviators — and of Mawson himself — the aircraft was not erected until 22 December, but was not flyable.

Campbell was concerned at the Moth's vulnerable position, as it was susceptible to the variable winds. He sought approval from McKenzie, the ship's second-in-command, to lash it to the deck grating. Permission was granted, with the proviso that no rope be cut in the process, with the result that when the task was completed 'there appeared to be more rope than plane'. (5) For further security, it was decided that the floats should be removed and the aircraft bolted to the grate.

Problems with the Moth had to be solved before it flew on 31 December. The engine would not start. Constant swinging of the propeller brought no response save for spasmodic engine backfires, and numerous experiments were undertaken in attempts to make it run. Campbell recalls that two strip heaters were fixed to the crankcase and the engine primed with medical ether in one attempt to start the rebellious motor. (6) The initial problem was caused by the magneto insulation having been soaked, and thus swelling. It had to be baked to make it serviceable, and Frank Hurley's attempt to heat the engine by placing Doctor Ingram's petrol primus on the tray underneath it almost ended in disaster. An unexpected roll of the *Discovery* capsized the primus, which exploded and burst into flame. Finally, Campbell and Douglas solved the problem by improvising 'a funnel of canvas . . . made with one end arranged to fit over the engine, with the other end . . . fastened against the ship's funnel within the fiddley. An electric fan was placed at that end to force hot air through the canvas funnel to the engine'. (7) To complete the process, the engine oil was heated on the galley stove before being transfered to the motor. After circulation of warm air for a night, the engine roared appreciatively at the first swing of the propeller, before it was stopped and closely covered to retain its warmth.

The last day of January 1929 marked the first flight of the expedition's aircraft. Lowering the aircraft overboard was difficult due to the ship's shrouds. Campbell later wrote:

> The seaplane took off and flew splendidly and was climbed in the vicinity of 6,000 feet. To the south lay the unbroken pack stretching as far as the eye could reach, its monotonous flat surface broken here and there by the towering mass of some imprisoned iceberg. To the south-west were darker shapes standing out boldly against the deadly whiteness. They were too big for icebergs — islands probably, or perhaps rocky outcrops on the coast itself. Their nature cannot, of course, be definitely established at this great distance, but it is well worth reporting as 'Appearance of Land'. (8)

Mawson named the land Douglas Island. He was convinced that the mainland had been sighted, and named the area MacRobertson Land, in honour of Sir Macpherson Robertson, one of the expedition's patrons. Further investigations had to wait the New Year, celebrated with a cabaret at which one item was 'On the *Discovery*', sung with gusto to the tune of 'Viva la Compagnie'. The relevant reprise for the intrepid airmen was as follows:

> We carry an aeroplane down all the way on the *Discovery*
> and two jolly airmen on Government pay on the *Discovery*
> they work winches and launches and hammer and screw,
> for the Moth she was stubborn and projects look blue,
> but now its New Year
> They've been in the air on the *Discovery*.

It was not possible to remove the floats from the Moth and during a severe blizzard on 6 January 1930 extra lashings were placed on the aircraft to guarantee its security. The two airmen watched over their charge as the ropes strained almost to breaking point. It appeared that they may break and the aircraft be lost overboard. Dagger shaped icicles, impelled by the howling wind, penetrated both the protective hessian covering the wing and the wing fabric. When the blizzard abated, the floats were removed and the Moth bolted securely into position — a hazardous operation in the prevailing weather conditions.

On 13 January 1930, at Proclamation Island, Mawson claimed the land between 73 degrees east longitude and 47 degrees east longitude in the name of King George V. After the ceremony, the *Discovery* began an attempt to survey the Enderby Land area. Part of the plan was for the Moth to undertake an aerial reconnaissance of the rocky mountains on the south-west coast.

In the event, these flights were not made, and ice conditions forced the expedition back to the Proclamation Island area, where the aircraft flew again on the 25th. Three flights were made that day. The second of these was to enable Frank Hurley to expose still and movie film. From a height of 2,500 feet, open water was observed to the lee of Proclamation Island. The third flight was flown by Campbell and Mawson, who was anxious to see the peaks which had been previously observed. When overflying the land ice, a flag was dropped with all due ceremony.

Despite a float being damaged as the aircraft was lowered into the water, Hurley and Douglas flew another photographic sortie on the following day. Further extensions of the mountain range sighted on the previous flights were noted. The mountains appeared to reach about 7,000 feet — some 3,000 feet above the altitude at which the Moth was flying. It was not, however, all work and no play, for the *Discovery* met the Norwegian antarctic expedition vessel *Norvegia*, and notes were exchanged. It was a surprised steward who welcomed Captains Holm and Rubbersen to the Mess, and offered the ship's hospitality to these two 'Norwegian' airmen. That no one had seen them arrive was immaterial, and a clean shaven and suitably attired Campbell and Fletcher were shown over the ship until they decided that the joke had gone far enough. [(10)] Mawson would have preferred to have stayed in continental waters for a longer period, but the opinion of the master of the *Discovery*, Davis, prevailed and the vessel set course for Kerguelen. Davis reasoned that his coal reserves were dangerously low, and course was set on 30 January. The wings and floats were removed from the Moth prior to its being bolted down for the passage. Campbell acted as a fireman for the trip; one of the firemen was incapacitated and in the doctor's care. The aircraft was reassembled at Kerguelen, where the airmen were involved in several incidents which were not related to aeronautics.

The scientists had taken over a hut at the end of the pier. As the island had become overrun by rabbits and large Norwegian rats, Fletcher describes one of the sports:

> From the windows of the room, rabbits and occasional giant sized Norwegian rats could be seen running about in the area between the

> buildings. Attempts were made to shoot them but were discontinued when Campbell, when rounding a corner, had to duck for cover as shots rang out. [11]

But some sort of revenge was in the offing. Both Campbell and Douglas were in a party led by Mawson on a two day field trip to the east coast. Again, Fletcher takes up the story:

> Earlier, Campbell and Marr had shown more than ordinary interest in the position of my sleeping bag . . . near midnight, Campbell and Marr, bellowing like bull elephants, raided our tents . . . [12]

At 9.30am on 18 February the Moth flew again, with Douglas and Campbell test flying the aircraft before Hurley took his place for a photographic flight. However, the wind had increased after the first take-off, and Campbell, having to swing the propeller whilst astride the floats, had difficulty starting the engine. The aircraft swung broadside to the waves, before drifting down the channel to smoother water, from where a successful take-off was made. Hurley was ecstatic about the results of the flight, although he may not have enjoyed the down draft which dropped the aircraft 1,000 feet on one occasion. Fletcher saw at first hand the reason for Hurley's reaction when he flew with Douglas on the last flight of the day, which the pilot terminated with an aerobatics display over the whaling station. On landing it was found that it was not possible to tow the aircraft back to the *Discovery*, so Campbell flew the Moth to the ship for recovery. On the following day Hurley joined Campbell for another photographic run, and it was decided that the pilots would be so employed while the ship remained at Kerguelen. The photographs were useful to the ship's officers in the compilation of charts. On one of these flights the airmen landed alongside a French sealing vessel anchored off Murray Island. Invited aboard, the Australians and Frenchmen toasted each other for some time — it was claimed 'to overcome the smell of seal oil and blubber'. [13]

The aircraft did not fly again after the *Discovery* departed from Kerguelen bound for Australia. On arrival at Port Philip on 8 April 1930, the ship was met by three aircraft from the RAAF base at Point Cook.

Mawson intended that, on his second BANZARE voyage, 'he would make considerable use of his little aeroplane to fill in geographical gaps, as it has a cruising range of 200 miles out and back, and from a height of 5,000 feet could observe outlets from the ice pack.' [14] The second expedition departed from Queens Wharf, Hobart, on 22 November 1930. Before sailing, a mysterious parcel was delivered to Stuart Campbell with an anonymous note requesting that the contents be taken 'south to ice covered Antarctica for (an) interesting experiment'. [15] The parcel contained an extremely well endowed brass monkey!

The *Discovery's* first port of call was Macquarie Island, where she arrived on 1 December. During the four days spent at the island by the expedition, Campbell assisted Hurley with his photographic assignments and Douglas was again employed as custodian of the launch. *Discovery* embarked from Macquarie Island on the 5th, with the intention of rendezvousing with the whale factory ship *Sir James Clark Ross*, with which arrangements had been made to replenish the *Discovery's* coal bunkers. Douglas was a member of the party which boarded the former vessel after the transfer had been completed. Another member was Hurley. Douglas was accidentally left aboard and took the opportunity of accepting the photographer's invitation to accompany him whilst filming from one of the whale chasers. Unfortunately a heavy dinghy had been rowed two miles to return the errant airman to his ship, and one can imagine that the crewman was not amused when Douglas and Hurley returned at 10pm, to a well earned reprimand from the expedition leader. The irascible Hurley claimed that the photographs were worth it.

Campbell and Douglas completed the erection of the seaplane on 21 December, in preparation for planned flights about Mawson's old camp site at Commonwealth Bay, on the western edge of King George V Land. En route to Commonwealth Bay, coal was again replenished, this time from the whaler *Kosmos*. After completion of this task, the *Discovery* was forced to seek the shelter of the pack ice as protection from a hurricane.

Claiming King George V Land and Oates Land as British territory gave the party the opportunity to carry out scientific observations, and also to acquire some new skills. Some members attempted to ski; others attempted to master a snow sledge. Among the latter were Fletcher and Campbell. Campbell's steering ability must be questioned, as on one run he, Fletcher and another found themselves heading for a recumbent Weddell seal. Hitting the seal amidships, the three men hurtled onto the animal's back. Initial displeasure at this intrusion was shown by the wildly floundering seal, but it promptly resumed its sleep. [16]

These diversions aside, the first flight of the series by the Moth was made on 7 January 1931, when Campbell and Douglas carried out an ice reconnaissance of some thirty minutes. The thick pack spotted forced the *Discovery* to proceed east-north east around its fringe. A similar flight was flown on the 15th, when Campbell and the cartographer, Oom, sighted land to the south-east at a range of some 95 miles. Another survey, flown next day, sighted land from an altitude of 8,000 feet, bearing south by east for a distance of about 110 miles, before tending south-west. The Moth had flown around 40 miles from the ship, which explains the extent of the sighting, which Mawson named 'Banzare Land'. Mawson and

Campbell attempted to fly to the area on the 18th at an altitude of 3,000 feet, but the area was obscured by cloud.

On 27 January, Mawson — anxious to spy the Antarctic continent — requested that the floatplane be made ready. In unfavourable weather conditions, Douglas and Mawson taxied down wind. The south-east wind caused a long ocean swell, which forced Douglas to curtail taxiing on several occasions. On the take-off run the aircraft was repeatedly thrown into the air without flying speed. Using all his skill, Douglas became airborne within 100 yards of the pack ice. Mawson gave a hearty 'well done' over the speaking tube.

The aircraft climbed to 6,000 feet where the clouds stretched to the horizon in all directions save to the south. A faint line showed in a sector of clear sky. Experience had shown that the continent was clear of cloud when the pack was shrouded so Douglas flew southward for some twenty minutes. As the aircraft turned to starboard, Mawson could see undulating ice-covered land, which he considered to be in the vicinity of Wilke's Knox Land. Three quarters of an hour later, Douglas throttled back and glided through the clouds, sighting the ship some four miles distant.

On landing, Mawson and Douglas attempted to hook onto the lifting shackle while the ship steamed slowly into the swell to decrease the vessel's roll and thus minimise any possible damage as the aircraft was lifted aboard. Douglas taxied the plane alongside, but the surge of the wash carried the aircraft out from the ship's side. The lifting hook was not engaged, and a second attempt brought no success.

On the third attempt, Mawson managed to hook the lifting sling over the hook. Douglas recounted that:

> The next instant the plane was lifted clear of the water with a jerk as the ship rolled and then suddenly the starboard wing of the plane went under the sea. Before we could appreciate what had caused this the plane tilted up vertically and Sir Douglas, who had been kneeling on the fuselage decking near the sling, fell toward the water but fortunately managed to grasp a strut and to hold on with his feet in the sea. I was still strapped in my cockpit but managed to release the strap and clamber up toward the nose of the plane in an effort to weigh the nose down and give some degree of righting to our machine.
>
> A few seconds later the plane's lifting sling broke and the aeroplane fell into the water with its tail and the rear part of the fuselage under the sea. We quickly dropped astern of the moving ship and could see and hear that consternation reigned as they made efforts to launch a boat to come to our rescue. During this time Sir Douglas and myself clambered to the bow of the plane's floats and, apart from our legs,

> escaped a ducking in the cold sea. Several minutes later the boat came alongside and took us aboard. (17)

The aircraft was repairable, the most damage being inflicted on the fuel tank. Hurley, Campbell and Douglas repaired the aircraft which undertook an ice reconnaissance on 6 February, albeit with a slight list to the right. The next day Mawson and Campbell flew from the lee of a large iceberg. From an altitude of 5,700 feet it was observed that the pack ice to the south-west was broken by lanes of open water. Ninety miles away to the south-east, the faint outline of land could be discerned.

Douglas and Oom participated in the final flight of the Moth under the auspices of BANZARE on the 11th. Taking off from a restricted area of water, the aircraft became airborne just prior to reaching brash ice. The aircraft flew to the eastern boundary of MacRobertson Land, Cape Darnley, and the crew sighted Cape Amery, Princess Elizabeth Land, and the western boundary of the Munro Kerr mountains.

Campbell was the first man ashore at Murray Monolith, when the land between 130 degrees and 60 degrees east of Greenwich, and south of latitude 64 degrees, was claimed for the King. He again featured at Cape Bruce, where he raised the RAAF flag, but had to return the champagne bottle which he souvenired as a memento of the occasion.

Discovery berthed at Hobart on 19 March, after a voyage home which was made memorable because of the weather encountered. At one stage the masts, some 90 feet tall, were actually below the crests of the waves. Mawson summed up the usefulness of the Moth in the following terms:

> The aeroplane proved a most important factor in the success of the geographical operations. That so much use was made of the machine, operating under difficult conditions, is owing to the determination and skill of the aviators, Campbell and Douglas, whose capacity is of the highest order. They deserve very great credit. (18)

VH-ULD was sold to the Australian Aero Club (Western Australian Division) on 24 July 1931, and was impressed into the RAAF as A7-94 on 22 July 1940. While operated by No.4 Service Flying Training School the aircraft crashed into the sea on 10 May 1942 and was converted to components. (19)

On 22 November 1935 the American explorer Lincoln Ellsworth and his pilot Hollick Kenyon departed from Dundee Island, 500 miles south of Cape Horn, to attempt a 2,200 mile trans Antarctic flight to the Bay of Whales in the Ross Sea. Eight hours later, wireless contact with the expedition's support vessel, *Wyatt Earp*, ceased. The expedition organizer, Sir Hubert Wilkins, assumed the worst. So did the Australian government. With the agreement of the British government, the Royal Research Society vessel *Discovery II* was placed at Australia's disposal. Although almost due south of Fremantle when the request for assistance was received [20] it was a fortnight before the ship berthed at Melbourne.

The Wapiti A5-37 is hoisted aboard Discovery II *prior to that ship's dash south to locate Lincoln Ellsworth. (Dept. of Defence Public Relations)*

Eric Douglas had been appointed to lead an RAAF search party consisting of Flying Officer Alister Murdoch (later Air Marshal Sir Alister Murdoch, KBE, CB, Chief of the Air Staff 1965-1969), Sergeants S.F. Spooner, E. Easterbrook, J.W. Reddrop, Corporal N.E. Cottee and Aircraftsman 1 C.W. Gibbs. The aircraft supplied by the Air Force was a De Havilland DH60G Gipsy Moth (A7-55) and a Wapiti MK1A (A5-37). Both aircraft were configured as float seaplanes, but were also provided with land undercarriages and skis. The skis were fitted by G.L. Grendon, after being manufactured in the workshop at Point Cook. [21] The Moth was fitted with an extra 12 gallon fuel tank, which would give an endurance of 4 hours 30 minutes at 80 mph. It was intended that this aircraft, due to its lack of communication facilities, would be used for local reconnaissance within visual range of the ship. The more powerful Wapiti was fitted with long and short wave transmitters with a respective range of 300 and 4,000 miles. It was also fitted with a Sperry directional gyro, sun and magnetic compasses and a pioneer drift indicator, and was thus more suitable for any long range search missions if these should prove necessary. [22]

To enable the aircraft to be stowed on the ship, structural modifications had to be approved by the Discovery Committee in London. The transom post was removed to allow the Wapiti to be stowed, [23] with the Moth being housed on the flat roof of the ship's hospital. The modifications to the vessel delayed the departure from Melbourne until 24 December.

The Gipsy Moth is lowered from Discovery II *prior to contacting the Lincoln Ellsworth expedition, 15 January 1936. (Author's collection)*

After averaging 220 miles per day, the crew of *Discovery II* sighted the mountainous coast of the south island of New Zealand, before landing at Dunedin on 31 December. The airmen stayed five miles from the town, but took the opportunity to complete the preparation of the aircraft. The ship, having refuelled to give it maximum steaming range, slipped moorings on 2 January 1936. The trip south was hastened by the news that 'Wilkins is speeding up on his way also to the Bay of Whales. But I (Douglas) think we will arrive first.' [24] The trip south had become a race between the two expeditions, and those aboard the *Discovery* could feel the headiness of the challenge — for it was, in all probability, one of life and death. By 7 January the airmen had completed all necessary tasks on the aircraft and personal gear. One thousand miles south of Dunedin, *Discovery II* crossed the international date line and entered loose pack ice.

Two days later the airmen prepared the Wapiti mainplanes for erection, moved the floats away from the fuselage and were looking toward two days of fine weather to rig the aircraft. The Moth had been partially assembled. At this stage the ship had penetrated 100 miles into the pack ice, which was formed by the prevailing south-east wind driving hummocky ice from the Bellinghausen Sea across to Cape Adare on the north-west tip of the ice free Ross Sea.

Good progress was made until the 10th. Heavy floes were being poled from the ship's rudder and propeller by all spare hands, and a reconnaissance flight was considered advisable to assist the captain in navigating *Discovery II* to open water. On the 12th, the Moth was launched from the side of the vessel when it reached an isolated pool of ice free water. At 10am Douglas taxied the floatplane to ensure that there was no ice in his path and took off, reaching an altitude of 1,200 feet. As he climbed into low cloud, Douglas observed better ice conditions 30 miles further south. The pool from which he had taken off had closed considerably during the duration of the flight, and Douglas was forced to evade ice on landing.

Low cloud and 20 foot thick pack ice encased the ship next day, making the ship difficult to navigate, and the ship's master, Captain Hill, was anxious that another ice reconnaissance be flown. The wind had picked up, the sea had become choppy, and ice obstructions made the attempt hazardous. Despite the danger, Douglas and Murdoch, after a long take-off run, climbed through cloud at 1,200 feet and flew south for 15 minutes. Clear sky and water were sighted 40 miles south of the ship. The recovery of the aircraft after the flight was hampered by the fact that *Discovery II* had the prevailing wind in the wrong quarter to sling the aircraft aboard. Douglas flew low over the ship, indicating this fact, and the vessel was turned about; even so, Douglas and Murdoch were unable to hook onto the hoist. Murdoch joined the crew of the motor boat which had been launched to assist, while Douglas ascended again. The aircraft landed and was recovered when more favourable conditions occurred.

The flight report was encouraging. It was evident that, after a few difficult miles, the ice conditions would become easier. Later that afternoon the ship gained speed and finally broke through the ice into the Ross Sea on 14 January, having traversed 380 miles of pack ice.

The delays had caused concern about the welfare of Ellsworth and Kenyon, but the advent of open water raised the spirits of the airmen.

As the ship approached the Ross Ice Barrier, on a heading for Little America — the base established by the American Admiral Byrd as a base for his flight over the South Pole — the airmen had their first sight of sea smoke and the unforgettable glare from the shelf. Reaching the Bay of Whales, two orange flags were sighted, fluttering on top of the Barrier to the east. It was known that Ellsworth's aircraft carried such signal strips. Half a dozen maroon signal rockets were fired, which exploded a thousand feet above the ship. Trepidation set in; no sign of life or movement could be seen on the Barrier ice. It was surmised that the two Antarctic flyers were dead or, more hopefully, were sheltering six miles south of the markers at Little America.

To ascertain that the castaways were at Little America, Douglas and Murdoch were lowered overboard in the Moth at 9.30 am on 15 January. If no sign of the airmen could be found, the plan was that the Wapiti would be used to undertake a thorough search of the possible track of Ellsworth's aircraft, the 'Polar Star'. The low sea temperature made a quick take off imperative, or the sea spray would freeze to the floats and the undersurface of the wings. A run of half a mile was necessary for the Moth to become airborne before it climbed slowly to an altitude of 1,000 feet. Levelling off, the aircraft was found to be decidedly tail heavy, probably due to water having frozen in the aft end of the floats during the climb. As they approached the Barrier, no horizon was visible due to the extreme glare from the ice merging with reflection from the cloud. Despite this yellowish glare, a few minutes later Murdoch observed:

> . . . what appeared to be black cracks in the ice below us but as we both looked we both saw the cracks 'stand up' until we could see they were poles rising out of the snow. We then realised that we were over Little America. I carefully circled the area and we then noticed orange coloured strips near the poles. Suddenly we saw the figure of a man appear as if out of a hole and he started to move his arms. This caused great excitement between us as we realised it must be either Ellsworth or Kenyon. (25)

A bag of food, cigarettes and a letter from the captain of *Discovery II* was dropped, the latter congratulating the survivors on their magnificent flight and suggesting that, if they were fit enough, they should go to the Barrier edge where they would be met by a party from the ship. Kenyon made the trek to the ice edge, arriving about midnight. Ellsworth 'had a chill and was remaining in his bunk but they were both O.K.' (26) A party from the ship attended Ellsworth at Little America during the following day, and returned to the ship with him.

Ellsworth sent a message of appreciation to the Australian prime minister on the 16th, and the party awaited the arrival of his ship, *Wyatt Earp*. This vessel joined the *Discovery* on the 19th. The aviators employed themselves by unshipping the rudder from the Moth to save undue strain on it and the aircraft's fuselage, before visiting the American base on 20 January.

The RAAF contingent aboard Discovery II *during the return voyage after rescuing Lincoln Ellsworth. Front l-r: Murdoch, Commander Hill (*Discovery II's *Master), Ellsworth, Douglas, Gibbs. Rear l-r: Easterbrook, Reddropp, Cottee, Spooner. (Author's collection)*

For the airmen the voyage home was frustrating, for they believed that they had achieved the purpose of the expedition. *Discovery* sailed to McMurdo Sound and the Balleny Islands (which had not been sighted since Scott observed them in 1904), making scientific observations en route. The Moth was flown on the 28th, when it made an ice reconnaissance. The airmen kept themselves active on aircraft maintenance, greasing parts to prevent corrosion, and general shipboard duties. During the voyage home, Ellsworth was presented with a model of the Moth, which had been constructed by Easterbrook and Douglas. They also inscribed the parachute which had dropped the original stores to Ellsworth and presented it to him. The inscription read:

> Compliments to Messrs Ellsworth and Kenyon from the personnel of the Royal Australian Air Force aboard the R.R.S. *Discovery II*, Bay of Whales, Ross Sea, Antarctica January 15 1936.

The only other incident involving Douglas occurred on 11 February, when in rough seas he 'was awakened by a tremendous lurch of the ship and gear in my cabin falling about. I got up and promptly charged the other side of the cabin, giving the top of my head a good bump.' (27) Douglas summed up the voyage to Melbourne by stating that, if he had the choice, he would not travel in a scientific ship again if he could avoid it. The atmosphere aboard *Discovery II* did not compare favourably with that aboard the BANZARE voyages in which he had taken part. The relief expedition berthed at Melbourne on 16 February with Douglas, for one, claiming, 'it's good to be back'. (28)

Douglas reached the rank of group captain in the RAAF, and died on 4 August 1970. Murdoch reached the pinnacle of his profession. Spooner and Cottee became squadron leaders, Easterbrook a wing commander and Reddrop a group captain. The subsequent career of Gibbs has not been documented. (29)

The RAAF experience of the Antarctic before the Second World War was confined to the BANZARE and Lincoln Ellsworth relief expeditions. In the former case the service supplied a commodity which was in short supply — experienced airmen. The aircraft had been supplied by Mawson and his backers, and there are precedents where the RAAF assisted other government and government-backed bodies in tasks which can only be described as being in the national interest. Mawson claimed British sovereignty over the Australian Antarctic Territory at seven separate sites. It is significant that the claim related to MacRobertson Land was based on aerial observation. In the case of the second experience, the RAAF was also following national aspirations. The fact that, in hindsight, Ellsworth and Hollick Kenyon were not in any real danger is irrelevant. At the time the expedition was mounted, there was a very real conception that the effort was a matter of life and death. It is to the credit of the RAAF that the equipment was made ready with such efficiency and promptness, and that the relief was undertaken in such a professional manner.

The Australian Antarctic Territory became the responsibility of the Australian government with the passing of the Antarctic Territory Acceptance Act, which came into force on 24 August 1936. With the responsibility came the requirement to undertake scientific endeavour in the region, and to this end Sir Douglas Mawson was moving to mount a new Australian Antarctic Expedition when the Second World War intervened.

In 1947 the government established the Australian National Antarctic

Research Expedition (ANARE), initially under the control of Group Captain Stewart Campbell, who had flown Mawson's Moth during the BANZARE voyages. In 1949 the organization came under the ambit of the Department of External Affairs as its Antarctic Division, thus indicating the importance of an Antarctic presence to Australia's perception of her place in the world.

CHAPTER TWO

Probing the continent

IN 1946 Stanley Brogden, the aviation correspondent, suggested in 'Aircraft' magazine that the RAAF should use its new Lincoln heavy bomber to attempt a trans-Antarctic flight. (30) Despite this proposal, it was not until the following year that the Air Force became involved in a less ambitious project.

On 13 March 1947, Group Captain Deryck Kingwell hauled a heavily laden Liberator (A72-386) from the airstrip at Pearce, Western Australia. After orbiting the ocean near Perth to check its radio altimeter, the aircraft set course to latitude 46 degrees south. At 11.20am, Kingwell commenced a reciprocal course, landing at Pearce after being airborne for thirteen hours. The crew of seven and a civilian meteorologist, Mr G. O'Mahoney, were exhausted after the flight. So too would have been the crew members of Catalina A24-352 which had become airborne from Crawley Bay an hour after the Liberator's departure as an air sea rescue vehicle. On the following day, another Liberator (A72-381) under the command of Flight Lieutenant C. Wright took off from Laverton, Victoria, to make a similar flight to a point 850 miles south of Melbourne.

Both these Liberator flights were undertaken to gain weather data for use on the next scheduled flight, which was intended to determine the suitability of a Lincoln aircraft for flight in southern latitudes, to collect meteorological data and to photograph Macquarie Island. The aircraft selected was A73-2, which was on issue to No.1 Air Performance Unit, based at Point Cook, Victoria. It had been modified for the task by the fitting of a second astrodome in lieu of the mid upper gun turret, a Lancastrian tail cone, radio compass, radio altimeter and two 400 gallon fuel tanks in the bomb bay. The flight to Macquarie Island on 15 March

Lincoln A73-2 which made the flight to Macquarie Island on 15 March 1947 and later during 1951. (RAAF Official)

was made at 10,000 feet, with favourable weather on the outward leg. At the island, despite the fact that the scope for photography was limited by low cloud, 239 oblique photographs were exposed. The aircraft completed the 2,600 mile flight in rapidly deteriorating weather conditions. The aircraft was airborne for 14 hours 35 minutes. For the crew of the Lincoln, it was heartening knowledge that a Catalina had been based at Hobart as an air sea rescue vehicle. (31)

These three flights had been undertaken under extremely adverse conditions, with high winds and low cloud being experienced by all crews. The meteorological data was highly prized by the weather forecasters, who were in favour of more regular flights.

In an interview broadcast by the ABC 'News Review' programme on the 15th, a Wing Commander McCormack, who planned the operation, stated '. . . in October 1944, the idea was born that it would be of some considerable advantage to Australia to gain some experience of flying conditions . . . in areas of low temperatures . . . (We) foresaw that aircraft could be of great aid for whaling, in assistance of ships, as cover for Antarctic expeditions. And we also wanted to see how aeroplanes would perform when flying in temperatures far below zero. Naturally, there was an enormous amount of research and organization to be done. For instance, we had to find out what effect magnetic variation would have on compasses, what type of weather would be encountered at various times of the year, and what aerodromes could best tackle the flight. We contacted as many authorities as we could find . . . we consulted American authorities, and Russian, and British . . . The weather bureau gave us

Hasselborough Bay, Macquarie Island, photographed from Lincoln A73-2 on 15 March 1947. The scientific station was later established on the isthmus. (RAAF Official)

detailed summaries of all information they had on the likely weather conditions, and it was decided that the flight would best be made in the summer months. When it came to the actual selection of the aircraft, we decided upon a Lincoln . . . with this flight we were not only able to make a complete pictorial survey of Macquarie Island but also find out how an Australian erected standard type of aircraft would behave in southern latitudes . . . the results achieved were in all ways completely satisfactory . . .'

This flight may be seen as an essential preamble to the activities of the subsequent ANARE voyages, which were scheduled to proceed south in December 1947. Under the leadership of Campbell, the executive officer of ANARE, a two pronged attack was to be made. The Navy had supplied the vessels for the use of the expedition, H.M.A.S. *LST 3501* (later renamed H.M.A.S. *Labuan*) and H.M.A.S. *Wyatt Earp*, (Lincoln Ellsworth's old depot ship). The former was to sail to Heard Island to establish a station and a reserve fuel depot for *Wyatt Earp*, and subsequently establish a similar scientific station at Macquarie Island. *Wyatt Earp*, in the meantime, would reconnoitre the Antarctic coast in the vicinity of King George V Land, with the aim of establishing a base on the continent. The ship would then return to Australia via Heard Island before the winter set in. The overall aim of the expedition was to carry out cosmic ray observations, geological surveys and operate weather stations which would transmit daily data to Australia. In addition, livestock experiments, to establish the suitability

Members of the crew of Wyatt Earp, *1948. Jones is second from the left and Gray second from the right (with RAAF windcheater). (Dept. of Information Neg. No. 497/20 via D. Vincent)*

Wyatt Earp *at the Continental ice edge, February 1948. Note the overhang of the Kingfisher's wings. (ANARE neg. No. 4883).*

of sheep and goats for the sub-Antarctic environment, would be undertaken.

To facilitate support for the expedition, the Secretary of the Department of External Affairs, J.W. Burton, wrote to the three Service Departments on 16 May 1947, stating that '. . . Group Captain Campbell will be required to secure quantities of general and specialist equipment and supplies. It would therefore be appreciated by the Executive Planning Committee for this expedition if you would afford any facilities at your disposal for Group Captain Campbell and enable him to obtain, as far as practicable, any items necessary for the expedition from supplies under your control.' (32)

The Department of Air agreed to act as a purchasing agent, subject to reimbursement from ANARE for the requisitions raised, on 13 June, and the functions of the Department were elaborated on 29 October. In this telex to the Department of External Affairs, the Secretary of the Department of Air, Mr M.C. Langslow, advised Burton that the following principles would apply:

a) Pay, allowances etc. of RAAF personnel will remain a final charge to the Air Vote.

b) Equipment and stores of types surplus to RAAF requirements will be made on loan or free issue. Kingfisher and Walrus aircraft and associated spares come within this category.

c) Funds of expedition will be charged with the costs of all equipment and stores specially purchased on behalf or issued from Air Force stocks and which must be replaced to meet RAAF normal requirements. On completion of expedition, its funds will be credited with value of items charged for but which may be returned to RAAF in fully serviceable condition. [33]

The implementation of these principles involved the resources of No.1 Stores Depot, Tottenham, Victoria. On 6 September 1948, the Chief Finance Officer, Department of Air, wrote to the Area Finance Officers in Melbourne, Sydney and Brisbane refining the procedures to be followed. In his letter he noted that all requisitions would be charged to the Antarctic Vote, and that all goods would be delivered to the ANARE store and claims rendered to the Secretary, ANARE, who would certify the claims and forward them to the Area Finance Office for payment of the contractor. [34]

The Kingfisher is swung out for its only flight over Antarctic waters, 22 February 1948. (ANARE neg. No. 41371B)

Consideration had been given to equipping the Antarctic Flight, which was established at Rathmines, New South Wales, with Noorduyn Norseman aircraft. In the event, a Kingfisher (A48-13) and a Walrus (HD874) were selected for the venture, and modified for the task. The Kingfisher was fitted with pre heating equipment originally fitted to the Martin Mariner flying boat, modified to fit under the engine cowling against the cylinders, and a quilted engine cover constructed to conserve heat. Tent type heating devices were designed to prevent the freezing of dope on application, heat being supplied by carbon filaments operated from the ship's 220 volt electrical supply. Both aircraft were painted yellow for safety reasons.

The crew of the Walrus, l-r: Meek, Swan, Dunlop, Smith, Short. (RAAF Museum)

The Walrus was allocated to the *LST*, where she could be secured fully assembled. Provision was made to secure the aircraft on the deck. The Walrus joined the *LST* on 15 November, prior to the ship proceeding to Melbourne for loading. While at Melbourne, the aircraft was stored at Point Cook. The RAAF personnel tasked with the flying and maintenance of the aircraft were Flight Lieutenant M.D. Smith, Warrant Officers G.C. Dunlop and P.G. Swan and Leading Aircraftmen B.G. Meek and C.E. Short. Arthur Scholes, a member of the scientific party, recalls the return of the aircraft to the ship:

> A yellow seaplane appeared as a speck out of the clear blue sky above Port Philip Bay on November 17. After circling the Navy landing ship for directions, the plane landed alongside us on the water. The Walrus was hoisted aboard by a derrick and swung to rest on a cradle forward of the bridge on the main deck. Cradle and plane were guyed down to the deck plates by heavy wire stays.' [35]

The first port of call for the expedition was Fremantle, where the Walrus was lowered overboard for a final test flight on 28 November. The flight lasted one hour and 50 minutes. The aircraft landed in the lee of Rottnest Island for recovery as the ship embarked on the voyage to Heard Island, where it arrived on 11 December. All members of the party assisted in the arduous task of unloading the expedition's stores and equipment.

It was not until the 13th that the Walrus, flown by Smith, made what proved to be its sole flight from Atlas Cove. Even though the weather was suitable for the flight it took some five hours to ready the aircraft. It was noted that cloud had increased about the northern portion of the island. Campbell states that:

> We finally got him off about 4pm with David Eastman to take movies, W/O Swan to take stills and W/O Dunlop to work the radio. He found clear weather down the eastern end of the island, and brought back some very good and interesting photographs and observations. From his altitude readings, it seems that Big Ben we have seen is a long ridge in the centre of the island, about 9,000 feet high with a flat top from which rises a small conical peak about another 2,000 feet. (36)

The flight lasted an hour and a half, and was recalled due to unsatisfactory radio communications and the lowering of the cloud ceiling to some 500 feet at the ship. A fog, rolling in from the north would lower the visibility even further. The Walrus was taxied onto the beach at Atlas Cove, where she was tied down to concrete blocks. For Smith and the crew it had been a satisfactory flight: as a result, Atlas Cove was decided upon as the landing and camp site, they had exposed the first aerial photographs of Big Ben, the mountain which dominated the island, and must have been anticipating further flights and successes.

Heard Island, 22 December 1948. 'The Walrus was on the beach looking like no Walrus should look.' (RAAF Museum)

These hopes were shattered when, during December 20 and 21, a gale gusting to 90mph thrashed the island. The members of the expedition were forced into shelter. Campbell, later walking along the bay, looked to the north and saw:

> . . . the Walrus which was on the beach looking like no Walrus should look. It had been left tied down and unattended on the beach and had apparently broken or worn through its ropes and been blown over in the storm. (It was later found that) the Walrus had blown over this morning probably when the wind swung to the S.W., but nobody had actually seen her go. She had blown over twice sideways, smashing up the 16 foot dinghy along side as she went, and was now lying on her side a total wreck. (37)

The loss of the aircraft was not critical, as it was assessed that the other site over which it may have been utilized, Cape Laurens, would be impossible to observe due to constant cloud cover. It was decided to salvage the radio and instruments. Campbell, in his report, stated that:

> The loss of the Walrus is a very minor tragedy. It made one flight here and has enabled us to obtain a better idea of the island than anyone else has ever had before, and although a few more flights would have been desirable, it has fulfilled its purpose. But it is a nuisance on board and interferes to a considerable extent with the handling of cargo through the hatch, and hence, as there is no vital requirement for it on the Macquarie Island trip, we do not propose to replace it. It can add little, except photographs, to our knowledge of the Cape Freshfield area. (38)

For the crew of the Kingfisher on the *Wyatt Earp*, Robin Gray and T.W. Lidell (aircrew) and R.D. Jones (fitter), the voyage did not gain even the modicum of success of the Heard Island expedition. *Wyatt Earp* sailed from Port Philip on 19 December 1947, after being dry docked for repairs at Williamstown Dockyard to remedy defects which had arisen during her voyage from Adelaide. Before sailing, Campbell had presented Gray with a tattered RAAF ensign, for which he had signed an affidavit swearing that it had been lost beyond recovery seventeen years earlier. (39) It was soon evident that the *Wyatt Earp* was not a suitable vessel for the task ahead. Bass Strait welcomed her with atrocious weather, and water poured into the aft accommodation area. The discomfort of the crew was not alleviated by the unreliability of the engines, which continually broke down. After being forced into Hobart, she sailed on Christmas Day for the Antarctic, only to be compelled to return. The ship sailed again on the following day, but conditions aboard were so severe that the Naval Board, fearful of its safety, ordered the vessel to return to Williamstown.

Heard Island as seen by the returning 1959–60 Antarctic Flight (Kevin Felton)

The Kingfisher had been partially disassembled by the removal of the floats and the tail assembly, to prevent it overhanging the bulwarks. The aircraft was re-erected and flown to Point Cook while the *Wyatt Earp* was made seaworthy. The vessel was declared ready after sea trials on 6 February 1948, and the Kingfisher was re-embarked. The ship sailed south from Melbourne on the 7th. Thirteen days later, after a most uncomfortable voyage, she crossed the Antarctic Circle. However, the delay in sailing meant that the *Wyatt Earp* could only approach the continent to within 30 miles of Cape Gray, in the vicinity of Commonwealth Bay, on the shores of King George V Land.

Due to high seas and the difficulty of launching the Kingfisher, the master of the *Wyatt Earp* was reluctant to employ the aircraft. However, the ship's first lieutenant and Gray prevailed, and the Kingfisher was assembled for flight. Although the afternoon of the 22nd was relatively calm, it was bitterly cold, and the erection of the floatplane took five hours to complete. It was not until the afternoon of 13 March that the sea and ice conditions were favourable, and aircraft was lowered overboard.

Even though the ship was situated in calm waters, it took an hour to complete the operation. Just after 1pm, Gray and Jones took off to seek an ice free lead for the ship to follow. With a cloud ceiling of 1,500 feet observation was limited; all that was visible was seemingly interminable icebergs, making it obvious that further progress toward the continent would be impractical. Gray landed the floatplane, and one more flight was made with the photographer, Laurence Le Guay, embarked for a short photographic mission. The flights had been in the vicinity of the Ninnis and Mertz Glacier tongues, and the recovery of the Kingfisher took three hours of extensive effort by the crew.

Wyatt Earp set course for Macquarie Island, where she berthed alongside *LST 3501* at Buckles Bay on 20 March. The crew and scientific members from the *LST* were employed in establishing the station on the island. *Wyatt Earp* Earp remained at Buckles Bay for three days, giving her crew the opportunity of using the hot shower facilities aboard the *LST*, before turning north for Melbourne. *Wyatt Earp* arrived on the 31st, and the Kingfisher was unloaded and flown to Point Cook. Later, the aircraft was taken on charge by the Department of External Affairs for use by ANARE. In the event, it was not called upon again.

Gray, in an article published in 'Aircraft' magazine, made the following observations about the Kingfisher:

> The aircraft was three times dismantled and re-assembled. She was constantly exposed to salt water and salt spray and when at sea she was drenched more often than she was dry. For almost two months she was smothered inside and out with snow and ice, usually solidly frozen. She was bumped and dropped and battered and generally ill treated under all sorts of difficult circumstances; yet in 55 hours flying, though not by all means all in ice, no major unserviceability was experienced. When the aircraft was assembled, the engine was run daily if weather permitted and when not moored on the water. Not even a spark plug was changed. [40]

In fact, the Kingfisher had flown less than four hours in the Antarctic.

The RAAF was called upon to fly to Macquarie Island later in the year. On 4 July 1948, C. Noble, the diesel engineer with the party on the island, was drowned while skiing over the thin ice of a frozen lake, and a replacement was imperative. Stuart Campbell made representations to the Air Force for assistance. Coincidentally, a Catalina amphibian (A24-104) was being used by the Aircraft Research and Development Unit at Point Cook for Jet Assisted Take Off (JATO) tests. This aircraft had the necessary range and capability to land at Buckles Bay, and the JATO facility would be most useful in the take-off from the restricted waters of the bay. After JATO tests at Point Cook, the Catalina flew to Hobart, from where the flight to Macquarie Island would commence. The crew was under the command of Gray, who acted as second pilot to Flight Lieutenant A.E. Delahunty. The replacement engineer was K.F. Keating, with Campbell scheduled to fly as a passenger. The first attempt was made on 25 July, but the mission had to be aborted due to a combination of bad weather and instrument failure. The unserviceabilities could not be remedied at Hobart, despite the use of Trans Australia Airline facilities, and on 27 July the aircraft was flown back to Point Cook in extremely bad weather. The Catalina returned to Hobart, after a fitter and spare parts had been flown from Rathmines to repair the aircraft, on 20 July.

It was not until 4 August that another successful attempt was made to fly to Macquarie Island. Campbell could not make the journey. On the flight from Hobart the weather steadily worsened, with the cloud base of 1,500–2,000 feet deteriorating to 500 feet and associated bad visibility. These conditions were prevalent when the island was sighted seven and a half hours after departure. The conditions at Macquarie Island were described in the report of the flight as:

> Overcast Cu (sic) and S Cu (sic), base about 500 feet, squally, intermittent drizzle with surface wind about 30 knots. The lower portions of the island only were visible, the higher part all being in cloud. There was a heavy westerly swell running which was curling round the north and south of the island giving a confused choppy swell on the eastern side of Buckles Bay with a moderate chop and a slight surf. Severe turbulence was experienced on the lee side of the island. (41)

The flight engineer, Warrant Officer Jack Vercoe, recalls the subsequent events:

> We taxied very close to a sandy beach where, with the aid of two anchors and running engines, we managed to hold the aircraft into wind while we unloaded Gray and the engineer (Mr T.F.Keating) into the waiting boat. Survival time in the sea at that particular time of the year was estimated at two minutes. In spite of the two anchors the aircraft would not hold in the prevailing wind. We recovered our anchors, allowed the aircraft to drift two miles out to sea, then taxied back to shore on one engine. This procedure was followed using alternate engines until Gray came aboard bringing with him first day covers of the first airmail to Macquarie Island. (42)

The aircraft used four JATO units to effect the take off. The return flight was diverted to the Royal New Zealand Air Force (RNZAF) base at Wigram near Christchurch, due to the prevailing westerly wind which made it impossible to reach Tasmania. During this operation (code named 'Sinbad'), Lincoln A73-15 had been maintained on search and rescue readiness at the RAAF base at East Sale.

Catalinas of the RAAF had not severed all links with Macquarie Island. On 5 January 1951, a request from ANARE was passed to the RAAF base at Rathmines, requesting that a Catalina be made available as an alternative for a QANTAS flying boat which was being tasked to evacuate Mr J.G. Windsor from the island. Mr Windsor's medical condition improved, and the Department of External Affairs advised that he had been evacuated by the French vessel *Commandant Charcot* on 26 January, and an aerial evacuation was not required.

RAAF Catalina at Buckles Bay, Macquarie Island, 4 August 1948. (ANARE neg. No. 1415)

Lincoln A73-2 also maintained links with Macquarie Island. The aircraft, captained by Flight lieutenant A.E. McKenzie, undertook a flight to drop medical supplies and fresh food to the ANARE personnel. Fitted with long range tanks, the Lincoln departed from East Sale at 12.30pm on 1 February 1951. The flight south was not without problems, as the aircraft suffered:

> Virtual simultaneous failure of all navigational aids early in the flight, coupled with the fact that the weather conditions varied so widely from the forecast as to practically preclude astro, and the HF/DF facility existed in name only on the outward flight. Basic manual DR Navigation was the only recourse. (43)

Contact was made with Macquarie Island at 6am under visibility conditions of one and a half to two miles. As the Lincoln manoeuvred into position to drop the stores, a 30 knot wind howled across the peninsula. The first storepedo overshot the landing zone, but:

> The remaining two storepedoes were dropped singly after one more run had been abandoned due to late corrections, and both fell within 20 yards of the target. The ground party advised shortly after that all storepedoes had been successfully retrieved and the stores were in good condition.[44]

The Lincoln returned to East Sale, having been airborne for 13 hours and 45 minutes.

There was a tragic epilogue which involved a member of the Heard Island Walrus crew. On 28 September 1948 Flight Lieutenant Smith was the captain of Catalina A24-381, which crashed and burnt at Lord Howe Island, with only two survivors. Smith was not one of them.[45]

Like the pre-war expeditions, the RAAF experience over the sub-Antarctic furthered Australia's national interests. It can be said that the use of service facilities and manpower was the most economical method available to the government of the day to gain a foothold in the territories which had been claimed by Sir Douglas Mawson during the BANZARE voyages of 1929–30 and 1930–31. The role of the RAAF was a supportive one, and the flights made were of both practical and humanitarian value. The photographs and data gleaned assisted in subsequent planning of the scientific exploitation of the sub-Antarctic islands and the Antarctic continent by Australian scientists.

There is no doubt that the RAAF would have been utilised to support the Antarctic Division's efforts to establish a permanent presence on the Antarctic continent if suitable shipping had been available for such an effort to be undertaken. It was not until 1953 that the *Kista Dan* was available for charter from the Danish J. Lauritzen Line, giving the Australians a ship capable of penetrating the pack ice, and making it possible to contemplate a permanent presence on the frozen continent. Part of that presence was the RAAF Antarctic Flight.

CHAPTER THREE

A permanent presence

IT was not until 1954 that the RAAF again prepared to wing over the Antarctic. On 4 January 1955, two Auster aircraft (A11–200 and A11–201) were loaded on the *Kista Dan* as the equipment of the Antarctic Flight. Commanded by D.W. Leckie, the Flight consisted of R. Seaver, F. Morgan and K.W. Duffell. The expedition personnel was farewelled by Mr R.G. Casey, the Minister for External Territories, from Port Phillip Bay at 4.30pm.

Auster A11–201 is towed from Point Cook to be embarked on Kista Dan. *(RAAF Official)*

The 14 day voyage to Heard Island allowed the aircrew and other personnel to experience the rhythm of expedition life. For Flight members, this consisted of taking part in the mess roster, helping cook, laying and waiting on tables coupled with the inevitable washing-up afterwards. On the second day out, the engine of A11-201 was run up and the oil drained, flushed with Arctic 80 oil, and replenished with the same lubricant. It was opportune that they did so, as the following day was the last fine weather experienced before reaching Heard Island on 19 January. Sergeant Duffell, who suffered considerably from seasickness, was bunked in the doctor's cabin, with Leckie and Seaver taking it in turns to bed down in the forecastle. At Heard Island, Duffell was one of the first ashore, to recover from his affliction.

Members of the Antarctic Flight 1954–55, l-r: Morgan, Duffell, Leckie and Saver. (ANARE neg. No.3228)

The engine of A11–201 had to be turned over regularly on the voyage; the propeller had been removed, so a cricket bat was used to rotate the engine. The aircraft was stowed on deck, and the close proximity of one of the weazels (a small tracked vehicle similar to a sno-cat) caused Leckie some anxiety. He feared that it would damage the front of the aircraft in heavy seas. To allay his fears, Leckie prevailed on the second mate of the ship to lash the weazel to a stanchion. Flying and immersion suits were checked for serviceability, and oblique photographs and maps of the intended area of operations studied. For relaxation, the ship's captain showed slides of Greenland and some on the making of the film 'The White South'.

Unloading at Heard Island took a day. For Leckie, it was not without incident. While evading the Fergusson tractor, he fell into the sea, to find that his immersion suit was of faulty manufacture, and did not seal about the top of his boots. Seaver, having cut his hand, was excused from manual labour on the second day. Problems precluded the return of the station's D4 tractor to Australia. It had been decided to attempt to load the tractor on a double pontoon for return to the ship, by building a platform onto which the vehicle would be driven. The pontoon would be placed alongside at high water, and the tractor transfered to it. In the event, the platform was too high and the receding tide did not co-operate, leaving the tractor still in an elevated position. A makeshift ramp had to be built to enable the tractor to be driven from the platform, which was tedious backbreaking work.

An attempt to land the pontoon with stores was thwarted by the weather, which became so bad that it was feared the ship may have to stand out to sea. By late afternoon, the weather had moderated to a degree which enabled the pontoon, cutter and launch to complete the backloading operation. The cutter was damaged, only being capable of turning to starboard. For Leckie, now thoroughly soaked from his day's exertions, the *Kista Dan's* engine room was a welcome site to thaw out. The work completed, the ship sailed for Kerguelen at 9pm.

Reaching their destination, the voyagers found traditional French hospitality and the opportunity for the airmen to carry out maintenance on their aircraft. Leckie was able to obtain eight drums of 130 octane fuel, which was mixed with the 80 octane carried for the Austers. Leckie and Seaver considered themselves most fortunate in not breaking a limb while handling water and fuel drums from a pontoon to the ship. Leckie also inspected the newly constructed airstrip on the island, which he considered suitable for Dakota operations.

The expedition, after an exchange of salutes, departed Kerguelen on 27 January. The first sighting of icebergs occurred on 31 January. Duffell was forced to clean water from the interior of A11-201. The F24 camera lens was also thoroughly cleaned, while Morgan spliced a bridle for use by the aircraft. The Antarctic Flight was now ready for action.

It was not until 2 February that the first flight took place. Just after completing the washing-up chores Leckie was requested by Phillip Law, the expedition commander, to undertake an ice reconnaissance. The *Kista Dan* had entered an ice free pool. The initial attempt at taking off was abortive — the water was so smooth that the suction of the floats could not be broken. Even running the motor boat ahead to break up the water was ineffective. Annoyed, Leckie taxied back along his take-off path at full throttle, swung around the edge of the pack ice, and took off across his own ripples. The motor boat was heading toward him and, as he hit it's bow wave, Leckie rocked the aircraft with the ailerons and 'it clicked onto the step with a good half mile of clear water ahead'. (46) At 1,500 feet the pilot checked his radio compass and VHF communications. To the south the Henderson, Masson, David and Casey ranges could be seen. Setting an approximate course to where the Mawson station was to be located, and memorising water leads, the Auster returned to the ship after examining the local terrain.

The second flight was aborted due to a snow storm, and the aircraft hoisted back on board. Law was able to brief the captain on the leads which had been sighted during the initial reconnaissance. The following morning the *Kista Dan* commenced the tedious push through the pack ice, which extended some 20 miles from the site selected at Horse Shoe Bay as the site for the Mawson station. The ice was some 30 inches thick, so it was obvious that skis would be necessary for continued aircraft operations. The Antarctic Flight members pooled their resources to fit A11-201 with this equipment, the aircraft being serviceable by midday.

A11-200 on the ice during the approach to the Mawson Base, February 1955. (ANARE neg. No.3221)

Three flight were made, followed by six more on 4 February. One of these was to enable Bob Dovers, who was to lead the winter party, to inspect the Mawson camp site. The tail wheel of the Auster was broken when a landing was made on very rough ice. Another flight entailed the transporting of the ship's captain on a reconnaissance of Horse Shoe Bay and ice leads, followed by a photographic sortie over the Henderson

Ranges. When returning to evacuate Dovers, Leckie pegged out a double landing strip, thus establishing the RAAF's most southerly landing ground.

The intensive flying operations brought strain on the Flight members. Every two and a half hours, the *Kista Dan* would stop her snail-like progress while the aircraft was refuelled. A long hose was draped over the deck from the Number Two hold to facilitate this process. Refuelling was not simplified by the penguins, which congregated about the ship in their hundreds. Morgan recalls that:

> Penguins managed to get under our feet at every step we took. They would solemnly and closely inspect everything we did, squawk excitedly to each other and then crowd in even closer. We learnt something from them, though. All of us had been troubled by having to work on slippery ice and we were all having nasty and painful falls. We noticed that the penguins moved around easily by taking short steps. When we followed their example, we found we too could walk without constantly falling over. (47)

Leckie and Seaver alternated between flying and acting as flight director, which meant neither gained relief from the pervasive cold. Acting as flight director entailed sitting for three hours with radio and binoculars, but this task was finally undertaken by scientific members of the expedition.

On 5 February, three weazels and two caravans were off loaded, to make the 17 mile run across the fast ice to Horse Shoe Bay. With them went the dogs, and both Austers were made ready. A11–200 was required for weazel escort duty and the other aircraft was prepared for scintillometer work. Leckie's diary entry for the day takes up the story:

> The day was perfect, an ideal day for aerial photography. Unfortunately, as it happened, the aircraft was required to escort weazels across the ice channels by the safest and quickest route to the mainland. I was called at 0400 hours and had a cup of cocoa and some toast. My take-off was delayed and I did not get airborne until 0545. My first job was to fly Mr Law to Mawson and to complete marking out a strip for the aircraft, so I took a dozen flags with me. We landed at 0620. (48)

Behind this laconic remark was drama, as two attempts had to be made to sight the flag markers which Leckie had previously laid. The landing was made on blue, pebbly, highly-polished ice, the skis creating no friction on the glassy surface. Law waited for the impact on the wall of an iceberg which was looming frighteningly close in the Auster's path. Leckie ground looped the aircraft, spinning it around and around, the extra friction on the edge of the skis bringing the aircraft to rest, a bare 30 yards from the face of the iceberg. Leckie calmly continued with the task of further

A11–201, fitted with skis, is prepared to support the Mawson Base approach party, February 1955. (RAAF Museum)

marking out the landing strip. [49] To protect himself from the intense katabatic wind off the plateau, Leckie was wearing heavy clothing. This, combined with the hard, slippery ice made the job very tiring. It took three hours to complete the task, and Law was worried as those aboard the ship had been advised that the two men would remain ashore for no longer than two hours.

A search was about to be commenced when the Auster returned to the *Kista Dan* and Seaver took off at 10.40am on weazel escort duties. When he broke off, the weazels had made good some nine miles. Leckie resumed the weazel escort duty at 2.30pm, after the Auster had been refuelled. When he returned to the party, he found them stuck at a bad channel near a big iceberg. Finding a route for the weazels a mile west of their current position, Leckie later reported:

> I tried to get them to take this route but they took no notice. I naturally supposed then that they had already inspected this crossing and found it unsuitable. I followed the crack along the iceberg where they were attempting to cross, eastward to see how far it extended. It went around in a slow arc until it faced north and opened out at the sea getting progressively worse. So I called the ship and gave the facts as I saw it and that the weazel sortie would have to be abandoned. I took a calculated risk at this stage and did a low run over the pack and landed

> alongside the weazel. I told Bob Dovers about the lead I saw and he asked me to take him up and show him, and to do a direct run to Mawson. This I did and he was most satisfied with the proposed crossing. (50)

The ventral tank was fitted to Leckie's aircraft, which took off again at 5.50pm. For the next two hours the aircraft guided the party toward the Mawson site, although it would be after dark when it was finally reached. They were going very slowly and marking their track with flags, and Dovers was instructed to proceed to the base site and stay there until the ship broke through the ice. Leckie landed near the weazels to report this to the party, and was asked to give a direct route to 'Horse Shoe', which was only about three miles away.

Sergeants Morgan and Duffell had been busy most of the morning getting A11-200 serviceable, having the aircraft ready to fly by 6.30pm. With both aircraft erected, it was only possible to stow one aboard *Kista Dan*. One aircraft had to be pegged down at Mawson. Leckie and Seaver bedded down the aircraft and flew the other Auster back to the ship at 10.35pm. Bad light prevented the pilots from giving further assistance to the weazel party, so both enjoyed a meal and a glass of sherry with the expedition leader. It had been a full day for all concerned, and the value of the little aeroplanes in guiding the party to Horse Shoe Bay was respected.

However, for the next two days, gale force winds and snow prevailed. The pressure of the pack ice forced the ship upward some two feet, and thin surface ice started to climb the port side of the vessel. The main anxiety for the aviators was the fate of the Auster which had been tethered at Mawson. In fact, the selection of the site could not have been better, as the aircraft required only minor repairs to make it fully serviceable.

On the 8th, even though the *Kista Dan* was still in the grip of the ice, 15 sorties were completed. The equipment transported included weazel rescue gear, kerosene, cooking stoves, gas cylinders, food and fuel. At the end of the day, both Austers were secured on Number Two hold aboard *Kista Dan*. One flight, made on the 11th, discovered a route for the vessel to enter Mawson Harbour.

The wind, however, had increased to gale force, and the *Kista Dan* was forced beam on to the wind, which gave no protection to the aircraft lashed on the hold. The aviators held fears for the aircraft's safety — well founded fears, as Leckie describes:

> . . . the officer of the watch came and woke me at 0400 and said that the aircraft were in danger. Sgt Morgan was first on the scene at 0405 and the remainder at 0410. The damage had been done and both

fuselages appeared to have been damaged beyond repair. Both the starboard mainplanes appeared damaged. Stabilizers and rudders were damaged. We were a sorry lot with the members of the crew trying to get 201 off 200. We had to use a block and tackle to bring 201 down against the wind. The position appeared to be hopeless...Sgt Morgan took out the scintillometer and radio compass in case the aircraft went over the side.

The captain came into our cabin this morning while the Antarctic Flight was commiserating over its loss. He mentioned that he was worried, as now there were no aircraft to lead him out of the pack. Mr Law approached us at breakfast time to see if it was possible to make one aircraft serviceable out of the two. We said that it looked hopeless but we would try. As soon as the wind abated we decided to offload the aircraft, clear the middle hatch of No. 2 hold, and put the aircraft back and try to make one aeroplane out of the two. We were not hopeful in any way but it was for the safety of the ship and the expedition, then we would do our best. I placed myself and Sgt Seaver under the direction of Sgt Morgan. If we could get a serviceable aircraft and a flyable one, all the credit would go to Sgt Morgan. His initial statement was that it was a six month job in an AD (Aircraft Depot). If we could pull it off, the age of miracles had not passed, also if the plane could be repaired it would not have flaps and the number of the aircraft would be 200 with many accessories from 201, including 201's starboard mainplane. The captain suggested it be called A200.5. (51)

The next day, 13 February, Law proclaimed the Mawson base for Australia, and Leckie was heartened when Morgan advised him that A11-200 could be made serviceable. On the 17th, the aircraft was test flown and found to be satisfactory, although landing and take-off runs were longer than before, due to the lack of flaps. For the Antarctic Flight, there was an incident which may have proved fatal. Duffell fell to the bottom of Number Two hold; he went over on his ankle on a piece of loose timber. Fortunately, his outflung hand grasped the rim of the hatch, not only breaking his fall, but also straightening him to enable him to land on his feet. Even so, he was placed under doctor's orders.

Flying the hybrid aircraft off skis continued until 22 February. The task became more difficult and exhausting, the aircraft having to be manhandled a mile and a half to the take-off point. By the time the three men had carried out this task, they were a lather of sweat. Unluckily, Morgan fell and hurt his arm, so the pilots were forced to operate without their ground support for a period. Duffell was fit enough to load camera magazines.

The fuselage of Auster A11–201 waiting shipment back to Australia. Components of this aircraft were used to rebuild A11–200 after the two Austers had been damaged aboard Kista Dan *on 12 February 1955. A11–201 returned to the Antarctic in 1959. (RAAF Official)*

With these difficulties, it was impossible to fly more than two sorties a day. Flying was intensely uncomfortable. Temperatures of 38 degrees Fahrenheit below zero at 8,000 feet caused Seaver to suffer from cramps. On one flight the engine failed, due to water freezing in the fuel lines. Leckie recorded:

> Force landed in a strong cross wind, across the bow of the ship and on rough ice, breaking the port main ski . . . the worst part . . . was dodging a small island. Ground looping helped. I fired another cartridge and as sufficient fuel had flowed through for a restart I slowed under power. Seaver said nothing but merely hunched in his seat waiting for the impact that did not come. (52)

The 30 hours flown in the rehabilitated aircraft proved the hardest the Flight had to fly.

Conditions had changed, so it was decided to fit floats to the Auster. Morgan commenced the tedious job on the 22nd. He was hindered by the winch jamming, and the fact that he had to work without the protection of gloves. Despite this, he completed the task by midday on the 23rd.

Kista Dan sailed from Horse Shoe Bay on the same day, leaving the winter party to its toil. An attempt was made to carry out a reconnaissance when the ship was in the vicinity of the Scullin Monolith, some 120 miles east of Mawson. Six attempts were made to fly the Auster, but ice collecting all over the aircraft made these attempts extremely hazardous. Leaving Scullin Monolith astern, the vessel voyaged toward the Mackenzie Sea and Vestfold Hills. Hundreds of icebergs barred the expedition from approaching the former, so the *Kista Dan* swung north on 26 February. By the 28th *Kista Dan* was outside the pack cruising to the south-west, passing icebergs, one of which was estimated as being six miles long. With the nights lengthening, the ship headed south into the pack ice at Prydz Bay, arriving off the Vestfold Hills on the evening of 1 March. The scene from the ship, which was anchored close to shore, was unforgettable:

> The evening was clear and the sunset was magnificent. The rays reflecting off the icebergs from the setting sun brought out many different colours. After the sun had set, the moon rose. But for the clear cold and icebergs standing out like dead ships, the scene could well have been a tropical one. (53)

A conference was called after dinner, and a photographic and scintillometer survey of the Hills requested. For the next two days, the *Kista Dan* was forced to keep her engines running as she dragged her three anchors in the bad weather. As soon as the weather lifted, the scintillometer was fitted to the aircraft. Only one sortie was planned — the captain wished to clear the area as quickly as possible. Leckie, attempting to start the engine on the deck before the Auster was lowered into the water on the 4th, had the misfortune to have the starter cartridge jam. Duffell repaired the damage, but misfortune again plagued the flight; Morgan released the aircraft prematurely with Leckie sitting across the engine cowling trying to release the hoisting bar with frozen fingers. The row boat was used in an attempt to tow the aircraft back from the ice floes, but the wind made the task impossible. Leckie started the motor and taxied back to the ship, where the aircraft was tethered and the recalcitrant lifting beam released with a pair of pliers.

With the ice encroaching into the selected take-off area, Leckie was forced to taxi to the west, while Seaver, the flight director, kept a lookout for ice. From near several grounded icebergs, the Auster had a reasonably good run between islands. If it did not become airborne, the take-off run would end with a collision with either new ice or an iceberg. For the watchers on the *Kista Dan* the take-off appeared to take an eternity. The Auster climbed laboriously to 5,000 feet. Leckie was impressed with the sight below him:

> The inland lakes were of a different colour from the ones near the coast. The former were a bright emerald green, the latter were the same colour as the sea, which is nearly black from the air. Through the centre of

this rock exposed area appeared to be a low ridge running south east-north west. The emerald green lakes inland were completely ice free. I could see the rocks around these lakes extend about a hundred yards into the water, which was crystal clear, from where it dropped away vertically. (54)

Leckie, having taken photographs, followed with a scintillometer run over the islands and the mainland before making a tentative landing back at the *Kista Dan*.

This was the last flight of the season for the Antarctic Flight. By 11pm a force 12 gale was blowing and the ship and crew spent a horrifying night dodging icebergs, and the ever moving ice. A victim of the gale was the Auster, as Law describes:

The aircraft certainly was a mess. Gusts of wind under the starboard wing had caused the main strut, connecting the port float to the fuselage, to collapse and the plane had lurched over on one side, crumpled and twisted, with the port wing tangled up in the lifeboat davits . . . (about) 0300 hours I watched dejectedly as another spar collapsed and the plane toppled forward onto its nose on the deck. The propeller was bent back under the fuselage and the starboard wing crumpled up in a horrible, grinding mess. Shortly afterwards, the wind blew the whole lot over the ship's side and the second mate cut the lashings to let the wreckage drift away. As he did so, petrol from the fuel tanks spilled over his hands and he writhed in agony as his fingers froze up. (55)

For 36 hours the *Kista Dan* drifted in the malevolent ice, the captain using the ship's engines to evade ice which appeared to threaten it. In the early hours of the morning the ship hit the ice, raising fears that it may have been holed. The impact showered the ship with ice, and it heeled 50 degrees. *Kista Dan* was allowed to drift back into the loose pack ice to lessen the spray from the furious seas. The spray was freezing in tons on the ship's upper works and threatening its stability. Drifting further into Prydz Bay, there was a real fear that the ship might not be able to be extracted, and an enforced winter's stay was a possibility. On the 11th, the ship finally freed itself, much to the delight of the crew and expedition members. It had been a harrowing experience.

Heard Island was reached on the 14th, after a rough passage. The *Kista Dan* only stayed long enough to embark returning personnel; the weather was still unfavourable. The next port of call was Kerguelen, where fresh water was obtained and the crew members entertained by the French.

Leckie used the voyage home to draft his report. He stated that the Antarctic Flight members were 'proud to have been chosen to represent the Air Force, in what must be the best equipped expedition ever to go to the Australian sector of the Antarctic'.[56]

Preparatory to the 1955–56 expedition to the base at Mawson, details of the RAAF involvement were discussed between members of the parties concerned. On 7 August 1955, the possible utilization of a Wirraway as a rescue aircraft was mooted. This proposition was rejected, as was the suggestion that a Wirraway engine be made available as a spare for the Beaver aircraft with which the expedition would be equipped. A further conference was advised that eight applications had been received from technicians volunteering to accompany the aircraft to Antarctica. From this number, two men would be selected. These men would be required to be capable of servicing the aircraft in fields outside the parameters laid down in the Engine and Airframe specialties.

Members of the Antarctic Flight 1955, l-r: Johanson, Seaton, Leckie and Sundberg. (RAAF Official)

The tasks of the Antarctic Flight were laid down as firstly the provision of short range air co-operation for the expedition and secondly, the study of the behaviour of service equipment under Polar conditions.[57] After training, Doug Leckie again led the Flight, consisting of John Seaton, G. Sundberg and G. Johansen to the south. The *Kista Dan* sailed from No. 2 wharf, Melbourne, on 27 December. The equipment which the Flight was to operate consisted of the surviving Auster and a Beaver. The former was stored in Number Two hold, and the Beaver was embarked on the hatch above. A screen, designed to prevent damage to the aircraft from ice falling from the ships rigging, was erected over the aircraft.

The Beaver is accepted by the RAAF. L-r: Leckie, Johanson, Sundberg, Air Marshal Sir J. McAuley, Mr R.G. Casey, Minister for External Affairs, unknown and Mr P.G. Law, Director Antarctic Division. (ANARE neg. No.4780)

Phillip Law, again in command of the expedition, had drawn up a programme designed to map much of the unknown coast of Wilkes Land, between Terre Adelie and the Shackleton Ice Shelf. To accomplish this, the *Kista Dan* sailed on course for Davis Bay. On 5 January 1956 the ice edge was encountered, and the Beaver flew a short reconnaissance to establish a route for the ship. Two days later the ship anchored close to Lewis Islet, enabling expedition members to land on the island and on the mainland opposite. Leckie and Seaton flew the Beaver fitted as a floatplane on a photographic sortie along the coast to Cape Bickerton in the east and Porpoise Bay in the west, a distance of 300 miles.

Leaving Lewis Island on the 9th, *Kista Dan* proceeded westward, and a further flight was undertaken which included both an ice reconnaissance and photographic duties. This did not establish a route through the pressure ice, but photographs of 30 miles of the coast west of Cape Southard were exposed. This was the last flight of the Beaver on floats. Not only did conditions become less suitable for using this aircraft in its float configuration, but on the 13th it was dropped from a height of 18 inches when a lifting sling broke. The Beaver dropped back onto its cradle, and the floats were badly damaged. Fortuitously it was possible to disembark the aircraft onto an ice floe and replace the battered floats with skis. The wheels of the wheel/ski combination had to be fitted aboard the ship. Three days later, the Beaver was flown with this combination. The flight confirmed that it was impossible for the ship to approach Cape Southard, so the expedition leader decided to head west to the Windmill Islands, making scientific observations at Thompson Island en route.

On the 21st, the expedition commenced a week's stay at the Russian base at Mirny. From there, the Beaver was employed to photograph the area east of the Bunger Hills, linking earlier photographic cover. After departing from Mirny, *Kista Dan* was beset by ice for 10 days, thus preventing a planned visit to the Vestfold Hills, before making landfall at Mawson on 17 February.

The changeover of equipment and men was completed by 4 March. Twenty men, including the four members of the Antarctic Flight, all under the command of William Bewsher of ANARE, remained to complete the programme. The airmen constructed a steel hangar to enable flying operations to continue into the winter months. The conditions were arduous. The men swung among the girders, some 20 feet above the ground, in 40mph winds to complete the construction.

The building shared by the officer-in-charge, geologist and surveyor, became the Antarctic Flight headquarters. A beacon transmitter and receiver was installed, and a beacon aerial attached to the radio mast. The latter operation required the assistance of all hands. Unfortunately, the aerial was not proof against a blizzard some nights later, and had to be re-secured.

On 27 April, Seaton attempted to douse a fire in the hut, but the fire extinguisher would not work. Others who attempted to quell the fire were almost suffocated by the smoke, and virtually blinded by the snow when seeking fresh air. The fire was extinguished without major damage to the building or its contents, although Seaton occupied much of his spare time re-lining, cleaning and re-painting the interior. [58]

The ice conditions at Horse Shoe Bay prevented any flying until 4 April, when the Auster was test flown. This flight was succeeded by ones to Stefansson Bay to search for sites to land field parties and note any points of interest. After two months of inactivity, the Beaver was flown on

20 April. On the 21st, Leckie and Kirby, one of the surveyors, flew the Beaver along a bearing of 225 degrees from Mawson. After flying some 300 miles, a nunatak — a mountain top protruding above the ice plateau — emerged from the haze to starboard. The pair sighted many more, indicating a mountain range. They estimated that the peaks reached 8,500 to 9,000 feet. Aware that any deviation from the flight plan would jeopardise any rescue attempts if a forced landing became necessary, Leckie flew parallel to the range. Kirby kept a check of their position with the astro compass, until they reached a position at 70 degrees south 52 degrees east. The next day Seaton flew 300 miles along the 63rd meridian, noting a continuation of the Prince Charles Mountains swinging gradually westward. On the instigation of the director of the Antarctic Division, Leckie and Kirby tried to carry out a further flight to the range that they had discovered, but bad weather forced them to abort the attempt. On the 31st, they made another attempt to obtain accurate data. While flying over the area at 1,500 feet, the horizontal visibility decreased, forcing Leckie to return to Mawson after being airborne for four hours. As he turned the aircraft, both pilot and passenger felt the symptoms of hypoxia. The aircraft descended through cloud, Kirby manning the radar altimeter, to emerge in a valley between the ranges. The Beaver landed back at Mawson after a flight of seven hours 45 minutes.

During May the Antarctic Flight undertook depot laying operations in the King Edward VIII Gulf area. Leckie flew the Auster to the area on 1 May to mark out suitable landing sites for the Beaver. The snow surface was too soft to enable that aircraft to land without skis. It was not until three days later that conditions improved enough to enable the Beaver to be flown to the depot site. Stocking the depot kept the Flight busy for eight days, but the flying task was interspersed with support of the field parties, which were undertaking observations in the area.

At the request of the two surveyors, Crohn and Kirby, air support was given to a geological and survey task at Stefansson and Scoresby Bay, even though there was only some five hours of daylight available each day. The operations were not without their difficulties, as Leckie discovered when he landed with one of the expedition members, McGregor, at the Crohn/Kirby camp at Stefansson Bay on 18 May. He reported that:

> On landing . . . and taxiing toward what appeared to be a lee behind a small rock monolith at the mouth of the Law Promontory, the winds appeared to strike the Auster from all points of the compass and the aircraft became unmanageable . . . (I) asked McGregor to climb out and to hold the wing of the Auster to assist with taxiing to a take-off position. It soon appeared obvious that more than one man was needed

> on the wing tips. McGregor's face was becoming rapidly frost bitten. (I) contacted Seaton (who) . . . said he would be over in a few minutes . . . (and) landed behind the Auster and McCarthy (another ANARE scientist) came over to assist the Auster crew. The wind at this stage was blowing off the plateau between 25 and 30 knots and bringing a reasonable amount of drift snow with it. At the camp site three winds met and the snow spiralled up vertically for several hundred feet. To save McCarthy and McGregor from taking too much punishment (I) decided on a short take-off and left McGregor and his equipment behind to return in the Beaver. [59]

The field party activities were completed on 25 May, when Crohn and Kirby were returned to Mawson.

Mid-winter operations were confined to routine test flights and essential maintenance of equipment. With longer periods of daylight in July, the Auster was utilized in laying food depots at Byrd Head and Foldoya and marking out safe Beaver landing strips at the same time.

John Seaton, Leckie's co-pilot, who overflew the Lambert Glacier on 28 November 1955. (Pix *magazine*)

Longer flights were attempted in August. On the 7th, Seaton and Sundberg discovered unmapped peaks in the interior of Enderby Land, and that Amundsen Bay was 'V' shaped and not 'U' as was indicated on the maps. Amundsen Bay was also encircled by mountains. Several islands rising 1,500 to 2,000 feet above sea level were noted, and a large glacier was observed flowing into the bay from the south-east.

Leckie, while undertaking a photographic sortie in the same area discovered a large bay adjoining the western tip of Amundsen Bay. This bay was 'U' shaped and rock-bound, but it was not until 11 October that the bay was further explored. Leckie flew to the area to duplicate a photographic run which had been unsuccessful, and noted a low-lying glacier running south between the Baggatt and Christensen Mountains. Radar altimeter readings indicated that the ice was no more than 50 feet above sea level, giving the impression that it was part of an ice shelf. Leckie assumed that the Christensen Range and the high ground to the north comprised a high island, and that the ice shelf extended to the west behind the land mass to join the sea in the vicinity of Kronsprons Olaf Kyst. The presence of large tabular icebergs lent weight to this theory. However, when a fuel depot was established at Amundsen Bay and Kronsprons Olaf Kyst was photographed, this theory proved erroneous.

Seaton, flying the Auster from King Edward VIII Gulf, force landed about 80 miles west of Mawson on 28 October. He reported that:

> After passing Cape Wilkins . . . I became aware there was no backward movement available in the elevator controls. This situation was quite tenable providing cruising power was used as the minimum power setting and no turbulence was encountered. I considered the position for some 15 minutes . . . and decided to attempt a landing in an area a few miles west of Taylor Glacier . . . A very long final approach had to be made . . . every time power was reduced the nose dropped and it was impossible to hold up due to the state of the elevator controls. The Auster touched down in the straight and level attitude at approx. 65 knots and only then was I able to pull off the power. A dead calm existed at the time and as a result the aircraft continued to run over the blue ice for what seemed miles; however the speed was dropping off enough to see that the end result would not be particularly hair-raising — even so, avoiding action had to be taken when approx. 150 yards from the glacier tongue and eventually the aircraft came to rest amid a rookery of Weddell seals . . .
>
> I found that the rear support of the radio compass control box had snapped, allowing the rear of the box to drop approx. two inches and control the control system situated beneath the instrument panel. I removed the control box and was able to take off and return to Mawson without further trouble. [(60)]

The work at Amundsen Bay and Enderby Land entailed expedition members undertaking astrofixes, magnetic observations, geological and biological experiments. These activities were supported by the Antarctic

Flight from fuel depots, and the parties were returned to Mawson after a week in the field.

On 4 November Leckie flew the Beaver 200 miles south into the Prince Charles Mountains to select a depot site to support the scientific effort. At the Stinear nunataks, he considered aborting the flight due to the pall of drift under which the nunataks hid. Bewsher and Crohn, who accompanied him, dissuaded him from this course of action. The drift was driving over the mountain peaks to form cumulus clouds, and the aircraft was tossed like a toy in the turbulence. Where the ice surface below appeared suitable for their purposes, the turbulence was so severe that the Beaver was almost uncontrollable, and where the air was quieter the ice was a jagged pattern of crevasses. Thirty miles south of Mount Bechervaise, a drift free zone was sighted in the lee of the mountain. Leckie selected a landing site on nobbly blue ice about a mile from pronounced medial moraine. He made a precautionary approach before touching down and, not wishing to risk the aircraft in the heavy drift and poor visibility, ordered the stores unloaded. Bewsher and Crohn remained behind to mark out a safe landing ground; a wise precaution, as the original site was found to pitted with crevasses. It was therefore essential that another landing site be flagged some distance away.

The site became known as Aerial Depot, and was the terminus of three tons of fuel, food and equipment, all of which was transported to the site in seven Beaver flights. Leckie describes the operation:

> During the whole operation, Seaton and the writer took off and landed in conditions of drift and high winds. At times the depot party could not be seen until they were almost alongside the aircraft. Landing was not difficult, however, even with the drift. In the beginning the marker flags were hard to pick up but, once the base began to build up, the strip was lined with 44 gallon drums, weazel tracks and spares, dog food, etc., making it easy to detect. In fact the aircraft could operate a regular schedule when all other forms of transport (tracked vehicles, dogs and even man hauling) were completely immobilised. [61]

With two members of the Antarctic Expedition as observers, Seaton undertook the last major Beaver flight on 28 November. The aim was to locate the extremity of the Prince Charles Mountains and the glacier which flowed between the adjoining mountain range. It was considered that the aeroplane would be in major jeopardy in the case of engine failure, as there was no landing area available on the tortured, crevassed ice. One of the passengers was landed at Aerial Depot to undertake magnetic observations, and the Beaver refuelled. Seaton tells the story of the flight:

The early part of the run took me over an area which I had already observed on a previous flight, but as we continued to the south east I could see a definite glacial stream bordered on its eastern edge by a continuous range of mountains and on its western edge by a group of block-type mountains protruding up to 5,000 feet above the ice. Other subsidiary streams joined the main glacier from the west, their paths clearly outlined as they made their way through these individual features. The prominent range on the eastern edge of the glacier had a well worn appearance; in various spots it was cleanly cut into sheer sided valleys by glacial action. Many short, steep gradient glaciers could be seen running into the main stream through these valleys. I noticed particularly the flat upper surfaces on these ranges and their cobble stone appearance — the deep valleys formed the western edge of the range into a series of bluffs. Behind the range to the east, the plateau had the appearance of being dammed up, its overflow of ice pouring through the valley toward the south. At the end of the range there was a great convergence of ice movement into the glacier.

At approximately 73 degrees 20 minutes south this mountain range petered out as the glacier swung toward the south east. The height of the glacier at this point was about 3,500 feet. To the west, the individual peaks of the bordering ranges were still appearing as far south as 74 degrees south, but beyond this only the eternal white of the plateau could be seen. I continued with the photography from the east side of the glacier to a position 73 03 south 68 east. In this position I was flying over the southern most peaks of the western edge of the glacier; to the north I could see the gentle sweep of the stream onto the eastern end of the Prince Charles Mountains and finally the Amery Ice Shelf. The small subsidiary glaciers flowing into the main stream from the west were very clearly outlined from this position; their stream lines and crevasses stood out distinctly.

At this stage, through fuel considerations, I was forced to set course for Aerial Depot again. On the return flight, as I was not taking trimetrogon photographs, I was able to alter course at will to study some of the western mountains and the nature of their terrain. They retained a similar appearance to those on the eastern side but were in individual positions; sheer sided and with the same cobble stone tops, they presented a spectacular sight. Three distinct tributaries of ice ran into the main glacier, their confluence causing high pressure ridges at their junctions and, in other areas, crevasses of considerable length which would form an impenetrable barrier to any form of land transport. The whole of this western area was the same, with the crevassing continuing well north into the Prince Charles Mountains.

Lambert Glacier, one of the largest in the world, over which Seaton flew on 28 November 1955. (ANARE neg. No. 9588A)

> From my observations on this flight, I would say that the glacier is roughly 200 miles in length, varying in width from 15 to 40 miles. At 73 south its height is approximately 3,200 feet and at 73 30 south in the vicinity of 3,000 feet. The mountains on the east are upwards of 6,000 feet while those to the west go to an estimated 8,000 feet. [(62)]

Seaton's flight was certainly the most spectacular of the season. He had flown over the Lambert Glacier, one of the largest (if not *the* largest) in the world. The subsequent three flights were anti-climactic.

CHAPTER FOUR

Consolidation at Mawson and Davis

ON 17 December 1956, the *Kista Dan* sailed from Melbourne carrying the Antarctic Flight led by Peter Clemence, and comprising D. Johnston, R. Pickering and N. Merideth. A Beaver, supplementing the Beaver and the Auster already at Mawson, was part of the ship's cargo. The intention was for the *Kista Dan* to sail to Mawson, undertake relief operations, and sail 400 miles to the east, where a base would be established in the Vestfold Hills. A rendezvous would then be made with the U.S.S. *Glacier*, to pick up members of the Antarctic Division who could not, for space reasons, take passage on the Australian vessel.

On 4 January 1957, the Beaver flew from an ice pool in the pack ice. As observed from the aircraft, the pack ice covered 70 miles. It was impossible to penetrate to the Mawson station, so it was decided to undertake a reconnaissance of Prydz Bay and the Vestfold Hills. The aim of the flight was to examine the seaward approaches to the area, but the airmen's view of the land was obscured. After the three and a half hour flight, the airmen considered that the Vestfold Hills were also enclosed by fast ice; however, a large unmapped rock 30 miles to the north-east of the coastline was ice free. These sightings were later proved to be incorrect and caused by a map reading error; the rock sighted was, in fact, the Vestfold Hills and the Larsemann Hills had been mistaken for them. A second flight was made toward Mawson in an effort to find a route along the coastline for the ship, but low cloud impaired visibility. The Beaver flew to Mawson on the following day, dropping much awaited mail.

It was decided to establish the base in the Vestfold Hills before the relief of Mawson was undertaken. On 9 January, *Kista Dan* was 120 miles from the Hills in comparatively ice free water thus enabling an ice reconnaissance to be flown. More flights were made on the following day to discover a suitable site for the station, one which had a deep water anchorage adjacent for the ship. Once the site had been selected seven days of construction work and the landing of stores was required to establish the station, which was named 'Davis', as a memorial to Captain J.K. Davis who had been captain of the *Discovery* in which Mawson sailed in 1929, and one of the pioneer navigators closely involved with Australian Antarctic exploration. The Beaver flew trimetrogon photographic sorties over the coastline from the Western Ice Shelf to the western extremity of the Amery Ice Shelf, and undertook vertical photography of the Vestfold Hills during this period. After the naming ceremony on 13 January, the *Kista Dan* sailed for the Windmill Islands, where she rendezvoused with the U.S.S. *Glacier* on the 25th. The entry into Mawson was delayed by three days of inclement weather. This did, however, clear the ice from Horse Shoe Bay, enabling the ship to enter the harbour on 3 February. It took two weeks to complete the changeover operations. During the period, the wings of Beaver A95-201 were removed and stored along the hangar wall, to enable both Beavers to be stored in the protection of the building. Flights were made along the coast to King Edward VIII Gulf, 140 miles to the west, to enable the new Flight members to become familiar with landing areas, landmarks and depot sites.

Members of the Antarctic Flight 1955-56. L-r: Clemence, Johnston, Pickering and Meredith. (RAAF Official)

Kista Dan departed for Melbourne on 17 February. It was planned that the aircraft would be involved in the support of field parties in the Prydz Bay, Enderby Land, Prince Charles Mountains and Foldoya areas throughout the year. Flights would be made to Davis to enable the rotation of personnel. To undertake these tasks, depots would need to be established at Beaver Lake and Casey Bay.

The programme could not commence until the sea ice had reached a sufficient depth to allow the aircraft to operate from it. The first flight of the Auster was on 17 April, with the Beavers being ice bound until the 25th. Ten flights had been made in preparation for the task ahead by the end of the month. A problem which was to plague the aerial operations for the whole season manifested itself: the base radio equipment was below standard for aircraft use, with extreme noise levels making radio communications uncertain. Even so, a flight was made to King Edward VIII Gulf on the 30th to enable the study of an Emperor Penguin rookery, and enable the depots established during 1956 to be inspected.

The first flight to Davis took place on 1 May. Flying via the Larsemann Hills to avoid open water in the McKenzie and Prydz Bay areas, the aircraft flew through overcast conditions, giving the crew little view of the terrain over which they flew. The conditions at Davis were clear. The cargo — a sled dog, radio operator and surveyor — was successfully delivered. On the following day, the aircraft was airborne over the Vestfold Hills, seeking flat rock areas which would give a sledding party access between adjacent fjords. The aircraft returned to Mawson on the 3rd, succumbing to the prevalent radio problem en route.

The Auster had been flown to the Taylor Glacier with a meteorologist and radio operator in the absence of the Beaver, to enable biological studies to be made of the Emperor Penguins. The Beaver returned to Davis on the 8th, with an auroral physicist, a geophysicist and another sled dog as passengers. The physicist was to establish an 'all sky' camera, while the geophysicists took gravity measurements at the station. On the return trip four days later, the aircraft experienced airframe icing at 6,000 feet. The Beaver was subsequently utilized to transport the Taylor Glacier party to Foldoya. After unloading, the opportunity was taken to seek landing sites at Steffanson Bay.

Another flight was attempted to Davis on 16 May. However, overcast conditions were met over McKenzie Bay, and climbing the aircraft to 14,000 feet did not clear the gloom. The flight was aborted when Davis reported that the weather there had deteriorated to a low overcast and drift snow. A landing would have been impossible. Another attempt was made on the following day:

> After passing the nunatak, Kjerka, low overcast cloud obscured the plateau. The flight continued above cloud and . . . at ETA Larsemann Hills, heading was altered as per flight plan for Davis. The planned heading was 114 grid and the expected time for the leg 34 minutes. The radio beacon at Davis was requested and it indicated that a heading of 090 grid was necessary to reach Davis. This heading was flown for 53 minutes before an overhead indication was received. This showed that a track error of 30 miles had existed at the turning point, had the beacon or radio compass failed, it is doubtful if the aircraft would have reached its destination where the cloud base had lowered to 800 feet and visibility to 2 miles. This deterioration had taken place in the last hour and no advice had been passed to the aircraft. [63]

The aircraft returned to Mawson via Mount Caroline Mihkelson, at the southern edge of Prydz Bay, on 19 May. This track was selected to establish, with certainty, whether the mountain was volcanic or not.

Construction of the base at Davis. (RAAF Official)

The aircraft only flew short range flights during the June winter conditions. The opportunity was taken by the fitters to give the Beaver a 100 hour service. With the winds now racking the base at 60mph, the hangar would have been almost comfortable. A three day blizzard, starting on the 5th, and bringing wind gusts of 110 mph, destroyed portion of the radio physics laboratory. In the aftermath, it took three days to clear the drift snow from the inside of the hangar.

On 11 June, a two man biological team was flown to Foldoya. As the camp gear which had been left by a previous party had vanished, the party was flown to the Taylor Glacier rookery. On the 14th they were flown to Foldoya, where a new camp was established. The landing sites which had been discovered at Steffanson Bay were utilized and found to be of an ideal surface for the Beaver.

During a flight to Cape Bruce on 3 July the expedition members came face to face with history. The aim of the exercise was to measure the movements of the Taylor Glacier. While undertaking these readings, the proclamation cairn built by Sir Douglas Mawson so many years previously was rediscovered. Two days later a depot, which had been restocked by the Russians at King Edward VIII Gulf was checked and gravity readings taken.

A flight to the vicinity of the Scullin Monolith and adjacent Mount Marsden to attempt to discover landing sites was inconclusive, due to white out and extreme turbulence near the Monolith.

On 20 July the Beaver and the Auster flew to the vicinity of Mount Henderson, in an attempt to locate a site which had been occupied by a field party during April. The camp site, 18 miles from the south-east of the mountain, had been marked by small red flags, which proved inadequate for the task. After tracking eight miles from a nearby rock outcrop, the airmen found the camp after considerable effort. It was decided to use more flags and spread four gallons of oil over the site to make it more prominent from the air. As the camp was 3,000 feet above sea level, the Auster was near its altitude limit for take-off and required an inordinately long run to become airborne.

The Beaver returned to the site three days later. In a 40 knot wind, it gave an impersonation of a helicopter during landing and take-off. The landing site was still difficult to spot from the air, so it was arranged that the field party would fire a Very cartridge at a pre-arranged time on the pick up day as a guide for the aircraft. This system proved successful: later tests proved that a red Very cartridge could be seen from a range of 30 miles.

Another attempt was made to plot landing sites near Murray Monolith on 6 August, but no satisfactory landing area was spotted. The return flight was made via the Douglas Islands, where a large Emperor Penguin rookery was discovered ten miles from the coast. It was estimated to have a population of 12,000 birds, making it one of the largest known rookeries in Antarctica.

The last flight of the winter programme was a gravity survey between

Mawson, Douglas Island, Oldham Island and the Taylor Glacier. The aircrew were concerned that the eight take-offs and landings which were involved would cause the gravity meter to 'drift', but their fears were without foundation.

August brought welcome flying conditions. In the first half of the month, the aircraft were utilized for 12 consecutive days. On the 9th, a surveyor and his field equipment was flown to Davis to commence duty on a geological survey of Prydz Bay. Camps had been established in the Larsemann Hills and the southern end of Sandfyord Bay. It was planned that the aircraft would track over these camps while en route to Davis. Excessive fuel consumption was noted while the aircraft was flying over the western edge of the Amery Ice Shelf, forcing the aircraft to divert to Davis where it landed with only six gallons remaining in its tanks. The engine was checked by the fitter who had accompanied the flight, and a subsequent test flight recorded normal fuel consumption figures.

Bad weather delayed the flight to the field party in the Larsemann Hills until the 12th. The aircraft landed a Sigdoy Islet, only 400 yards from the campsite. The return flight to Mawson was made on the 13th. Entering cloud just after leaving McKenzie Bay, the Beaver was still in cloud when the pilot requested the use of the radio beacon from a range of 20 miles. A fortuitous break in the cloud enabled the crew to identify the position of the aircraft — 25 miles south of the station. The radio compass indicated that the station was still further west. Again, excessive fuel consumption was recorded during the flight. The carburettor was changed, and the subsequent air test returned acceptable fuel consumption figures.

It was planned to fly to the Russian Base at Mirny. An attempt was made on 16 August, but:

> At Kjerka, although the aircraft was 3,000 feet above the plateau, severe turbulence was encountered with vertical air currents in excess of 1,000 feet per minute in both directions and swirling clouds of drift snow were carried aloft to two or three thousand feet above the surface. By the time McKenzie Bay was reached, cloud build up from ground level to 14,000 feet and advice that Davis weather had deteriorated to low overcast and visibility of two miles, caused the aircraft to turn back for Mawson. (64)

This foul weather over the McKenzie Bay area made the trips to Davis hazardous. On the 23rd, an attempt had to be aborted due to a combination of severe turbulence, cloud and snow showers, which extended from sea level to a height of 12,000 feet over the Bay. Another attempt on the following day resulted in the Beaver being diverted to the Larsemann Hills.

Weather conditions had forced the cancellation of the Prydz Bay ground survey, and it was decided to return the men and equipment back to Davis. The quantity of stores to be transported required two Beaver flights with a night landing probable at the end of the second. This had been anticipated by the expedition members at Davis, who established a human torch lit flare path — 'a typical example of the enthusiastic co-operation and invaluable assistance rendered to the Antarctic Flight throughout the six months that aircraft were operated in their region.' [65]

The field party's visit to Sandfyord Bay had to be cancelled, but the presence of the aircraft gave the scientists the means of gaining data of some value. The aircraft flew a geologist, a geophysicist and a surveyor to the north of Mount Caroline Mihkelson, where readings and samples were taken. On the return flight, the Beaver landed at the uncharted Lichen Islet, and an Emperor Penguin rookery was discovered at Flatnes Glacier.

On 4 September the long awaited flight to Mirny was undertaken. Due to fuel restrictions, only the one flight could be made. After flying for four hours the Australian aircraft landed at Mirny — to the surprise of the Russians, who were not expecting the visit. The Beaver crew were entertained with traditional Russian and Antarctic hospitality. The IL12 transports and MIL2 helicopters present were flown by crews who had long experience of flying over the Siberian and Arctic wastes.

Work on assembling the second Beaver (A95-201) was completed on 10 September. After a test flight, the aircraft was deployed to Davis to assist with the move of men and equipment back to Mawson. The radio facilities at Mawson proved a problem when the two aircraft returned on the following day. An unpredicted north-west wind forced the aircraft to drift south of the planned track, and the weak signal from Mawson was not strong enough to trigger the aircraft's homing mechanism. A break in the overcast south of the Mawson Range enabled the flight to fix its position and let down visually. Within 30 minutes of landing, the visibility had deteriorated to 200 yards, which would have made a landing impossible.

The initial flights over the Enderby Land coastline were made on the 17th, covering all intended camps and landing sites at Magnet Bay, Proclamation Island, Mount Biscoe and Amundsen Bay. A preliminary reconnaissance of the Prince Charles Mountains was also undertaken. A landing was made at Beaver Lake, 220 miles south-east of Mawson, where it was planned to depot 500 gallons of fuel whilst the survey teams worked in Enderby Land. The lake, with the exception of the shoreline, was suitable for aircraft operations. Two days later, a radar heighting flight was flown to determine the southern boundary of the Amery Ice Shelf and then extending to the eastern boundary of the Prince Charles Mountains.

The first camp for the Enderby Land survey was created at Magnet Bay on 25 September. On the flight to the site an Emperor Penguin rookery was sighted at Kloa, on a rock peninsular five miles south of Cape Boothby. The planned move to Proclamation Island had to be curtailed; the sea ice had broken out since the last reconnaissance. Only a small fringe of ice surrounded the island, and the ice which connected it to the mainland appeared hazardous. The aircraft, not having enough fuel to transport the party to the alternative site, Mount Biscoe, returned to Mawson.

At the end of September, a route reconnaissance was flown to plot a path for a team which was to undertake seismic work to the south of Mawson. The flights found an almost crevasse free route, and proved that there would be no inordinate problems with finding a passage through a series of ice domes.

The sea ice along the Enderby Land coast was turning treacherous, so it was decided to utilize the aircraft to accelerate the progress of the survey of that coastline. The aircraft would fly a party to selected points where two-hourly observations were to be made. The landing sites at Proclamation Island and Mount Biscoe proved excellent. Snow dunes at Amundsen Bay forced the aircraft to land two miles from a small island near the camp.

On 9 October, a camp was established on the western segment of Casey Bay, near Mount Christensen. The landing site was covered with soft snow and high power was required during taxiing of the aircraft. A fuel depot and camp were established adjacent to the landing area, after which the aircraft returned to Mawson via the depot at King Edward VIII Gulf.

The first fuel dumping run had been flown to Beaver Lake and the Prince Charles Mountains. Heavy ice rafting along the lake perimeter made access to the shore difficult. Two flights had also been made to survey the route for the southern seismic party from Mawson, through the David and Masson Ranges, to within 20 miles of Mount Twintop. From Mount Twintop the airmen flew along a track of 180 degrees, discovering a second east-west line of ice domes ten miles south of those originally sighted earlier in September. The geographical sightings were of interest:

> From the position 73 latitude 62 longitude, a large mountain range was visible 25 miles farther south, tailing off to the east though linking with southern extremities of the Prince Charles Mountains. No rock was visible to the west, but low cloud and drift in that direction gave

little scope for observation. The highest peak was estimated at 12,000 feet and the area plateau height at 8,000 feet. A well defined glacier and associated crevasse zones ran east and west between the ranges and the flight terminal. Several nunataks were visible to the north and north west of the range. (66)

Beaver Lake was now a base from where the aircraft ranged to select camp sites on the northern side of Mount Johnston, and, 90 miles further south, Mount Stinear. It was noted that the Mawson escarpment was closer to the Prince Charles Mountains and shorter than had been previously charted. A flight was organised from the eastern boundary of the Prince Charles Mountains to the high peak which had been sighted on one of the earlier route reconnaissance flights. The aircraft refuelled at Beaver Lake on the outward flight, landing and taking off on a surface covered with sastrugi (frozen wind blown snow) and snow dunes. The prevailing strong winds again caused the aircraft to impersonate a helicopter, but no major geographical features were discovered on the flight; those mountains already sighted were estimated to reach an altitude of 11,200 feet.

The navigator of the southern seismic party was flown to Mount Twintop before a second reconnaissance of the Amery Ice Shelf was flown. Due to fuel shortages this had to be limited to an area north of Patrick Point. However, additional information as to the southern extent of the Shelf had been gathered, and a radar altimeter run was made along the Lambert Glacier from east of Patrick Point to Beaver Lake.

Beaver Lake was established as a manned camp for the Prince Charles Mountain survey on 15 November 1957. A flight was made to the tractor train of the southern seismic party, six miles west of Mount Twintop, with the intention of exchanging radio officers. The terrain and the high altitude forced the Auster to return to Mawson — the aircraft would have required an excessively long take-off run to become airborne. Three days later the Beaver dropped supplies to the party, 30 miles further south. A smooth landing area had been sighted before the drop was made, and it was suggested that the aircraft return with a new radio operator, but the offer was not take up.

On 20 November, the Prince Charles Mountain survey team moved to Mount Johnston. It had been intended to use both of the Beavers on the task, but one of the pilots was indisposed, forcing the move to be made in two stages. Fuel holdings at Beaver Lake were replenished en route. This was prophetic, as fuel consumption figures during the last flight were again excessive. It was discovered that this was due to the backfire pressure relief door in the induction trunk remaining open after the engine was started, due to the hinge spring having lost its tension.

With the advent of summer, a large 12 inch deep pool of water formed immediately in front of the hangar doors at Mawson. This was the only avenue for the aircraft to be moved from the hangar to the sea ice, and it was possible that these may have broken through the ice. To protect them, and allow for recovery if necessary, the aircraft were winched across the affected area. This hampered support of the Prince Charles Mountain party, which was recovered and returned to Mawson.

The hangar at Mawson. Note the upright 'boxing' kangaroo on the Beaver. (RAAF Official)

A pre-computed daylight astrofix was undertaken at Mount Stinear on 29 November to complete the 1957 survey of the area. On the return flight, the aircraft landed at Mount Johnston to collect field equipment which had been overlooked when the party withdrew on the 25th.

Although it was still possible to tow the aircraft from the hangar by using a weazel, the state of the sea ice at the hangar frontage made further field expeditions unwise. The harbour ice was still firm. The decision was made to fly until fuel stocks were exhausted, or the sea ice collapsed near the hangar. Three 8-hour trimetrogon photographic flights were made over the Prince Charles Mountains and Enderby Land. That flown on 17 December was at the behest of the Antarctic Division, which expressed doubt as to the accuracy of the chart of the south-west ranges discovered in 1956. An exploratory flight was made, via the Leckie Ranges, which did not justify the markings on the chart. It was considered that the Tula and Scott Mountains, which were similar in configuration to those marked, may have been mistaken as inland ranges. This flight could have had disastrous repercussions, as:

> Fuel consumption on the outward flight was unusually low, but as the consumption rate had fluctuated throughout the year, no particular notice was taken of this variation. At the turning point of the flight, the possibility of continuing on the same track for another hour was considered as the apparent fuel state was sufficient, but in view of the excellent visibility and the fact that it was not possible to advise Mawson of the change in flight plan (there having been no radio contact for over two hours) it was decided to adhere to the flight plan and heading was altered for Leckie Range. On arrival at base it was thought that the wing tanks were still full . . . but an inspection showed that they contained less than six gallons . . . [67]

This was the final flight of the year. Only 130 gallons of fuel remained at Mawson, and it was considered prudent to keep this in reserve for emergency use. Thus ended the 213 flights of the season.

The flights of the 1956–57 season had proved the efficiency of small transport aircraft and their value to the scientific efforts of the Australians. They were available to transport men and equipment rapidly from point to point, thus enabling much to be achieved without undue danger and tedious travel by ice bound means. The aircraft had proved reliable and there had been no incidents which would prejudice the use of this form of transport in the Antarctic environment. On the whole, the advantages to the Australian scientific effort, and the safety of the personnel concerned in this venture, made the presence of the Flight a vital one.

CHAPTER FIVE

Aerial alarms and Wilkes

IVAN Grove had travelled south with the 1956 relief expedition. Now he commanded the current Antarctic Flight, the personnel of which were H.O. Wilson, G.K. Downer, S.A. Manning, A.K. Richardson and O. Maguire. With Grove, they set sail from Melbourne on the *Thala Dan* on 3 January 1958. With the airmen was Beaver A95-203, to complement Beaver A95-202 which was already in use at Mawson.

The expedition's first iceberg was sighted on 9 January, two days before

Antarctic Flight pilots 1956–57, Grove and Wilson. (Pix *magazine)*

the ship anchored at Lewis Islet. Leaving three members to complete the establishment of an automatic weather station, the ship departed on the 12th with the intention of spending eight days surveying the coast east of the islet. Fast ice impaired the ship's progress and an ice reconnaissance flown by Grove and Wilson confirmed that the only way of extricating the vessel was by steering a reciprocal course.

Progress around the continent to Davis was interrupted by visits to the French base at Durmont d'Urville and the Russians at Mirny. At the former the expedition divided, half being entertained by the French and the other half being hosts aboard the *Thala Dan*. Fitted in among these calls was a flight by Wilson to the American base at Wilkes on 28 January. This was a round trip of 240 miles, undertaken to transport an Australian observer, Mr Donovan, back to the ship. Grove described the flight in a letter to his parents:

> At about 6.30pm Bill leapt off into the blind. We had very little idea of the weather between our position and the coast, and absolutely no idea of which way the wind would blow him; and since he wasn't carrying an observer, we had no way of finding out. Things went fairly smoothly until he hit the coast some 80 minutes later; but because of the low cloud around the coastline, and the glare from the setting sun, Bill had difficulty finding out where he was. However, the Americans eventually plotted him on their radar screen and were able to vector him in. As far as I was concerned, that was the worst of the trip over, and since I was able to obtain from the Americans the time he arrived at their station, I was able to calculate his route and thus the average wind strength and direction which affected him on that first leg. By applying this wind to the return journey, I would surely ensure a safe flight back to the ship. But I was wrong. Somewhere along the track the wind made a change through about 270 degrees causing the aircraft to be blown off course quite a long way. Fortunately though, we were able to switch on the radio beacon and bring him home that way. All this mucking about took up precious daylight; and when he did arrive overhead it was getting dark making the job of alighting on the sea a difficult one. Anyhow, at about 10.30 pm Bill and the aircraft were safely aboard again and we set off, this time toward Mirny, the Russian base. (68)

Mirny was reached on 1 February. The consequent lunch was 'made memorable by vodka, caviar and Russian passion for mid Victorian appointments'. (69) Alan Richardson recalls that the vodka flowed in every hut. When reciprocating the Russian hospitality the following day, the

Australians were amused at the Russians' wariness of Australian wine. [70] *Thala Dan* departed at 1pm, bound for the station at Davis where she arrived on the 4th. The Beaver was flown over the Vestfold Hills two days later, undertaking a photo mosaic of the area.

The ship arrived at Mawson on 10 February. Two days later, a blizzard with strength of 50–60 knots prevented exterior work. As soon as the weather permitted, the Flight commenced the construction of a two bay extension to the hangar and the erection of a Metters self-erecting mast to replace the inadequate MF/HF beacon transmitting aerial. The Auster and Beaver A95–201 were crated for return to Australia, and Beaver A95–203 converted to a wheel-ski configuration. The ship sailed on 28 February, leaving 28 men alone to face the Antarctic winter.

On Anzac Day, Beaver A95–202 was put to bed after a number of training flights, to enable Flight members to change into uniform for traditional celebrations. Wilson, who had been flying to test the recently installed beacon, sighted the specially modified ANARE Jeep immobilised on the plateau behind Mawson. A subsequent search found two of the occupants walking back to base; a third had already skied to the buildings. A95–202 was not to remain under cover for long, as Alan Richardson noted in his diary:

> (We were) about to make (our) way to the mess and received word that Fred Elliot was overdue from a walk to Mount Henderson. (The) aircraft was moved out again to carry out a search. Ivan and self took off just on sunset to make the search. We flew direct to Mt Henderson 10 miles from base and circled it twice. On the third circuit I spotted Fred making his way toward the base ½ mile from Mt Henderson. We turned for base and signalled the ground party that he was O.K. Doc Channon and Helmut Schafer continued to meet Fred. We returned to base and landed almost in darkness. Fred walked in to the mess at 2130 having missed Doc and Helmut in the dark. Doc and Helmut continued on to Henderson in the hope of meeting Fred by which time it was too cold to stop so they continued back to camp arriving at 0100 hours. The Doc had frostbite in the nose, cheek and fingers. [71]

Another search party had not returned by dawn, and Wilson discovered them stranded by drifting snow. The two men were air lifted back to Mawson, although the horizontal visibility of only a few yards made the flight difficult.

The blizzards and the failing light of April and May reduced the flying effort, although flights were made to the Taylor Glacier and Davis. Prior to one flight to Davis, the aircraft had difficulty in taking off due to slushy snow having entered the tail cone while the aircraft was taxiing.

The first major flight was not made until 17 July. Wilson flew Operation Alfresco 1, a reconnaissance between Mawson and King Edward VIII Gulf, to open the spring programme. The gulf was covered in deep drift snow. No landing sites could be identified on the gulf, but one was sighted from an altitude of 2,000 feet on the north side of the most southerly of the Oygarden Island group.

Grove undertook an airtest with the Beaver fitted with two Nansen dog sleds hung on the inboard bomb racks of A95-202. As expected, there was some degradation of performance, and he was forced to use more nose-up trim than was normal. Later he flew the other Beaver on a reconnaissance of the Masson and David Ranges to locate landing sites for use during the forthcoming summer operations.

The Beaver fitted with Nansen sleds. (RAAF Official)

On 4 August, Wilson flew the doctor to Davis. He had been unable to make the flight on the previous day due to severe headwinds. August also saw the commencement of depot stocking operations in the King Edward VIII and Oygarden island areas. It also brought the Flight close to tragedy.

On 15 August, Grove, with Manning and McLeod as passengers, commenced a flight to Davis in A95-203. Complete cloud cover and strong winds, coupled with a report of similar weather at Davis, forced the aircraft to turn back toward Mawson. At 8.50am the oil pressure gauge in the aircraft read zero. Grove turned toward the sea, with the intention of landing on the sea ice. The mechanical condition of the aircraft forced him to land in a heavily crevassed area of sloping ice about 800 yards inland, eight miles west of Scullin Monolith. Grove wrote in the subsequent report of the incident that:

> The surface wind was about 50 knots gusting to 60–65 knots from the south, and to prevent the wheels falling into crevasses, the landing was made using skis facing up the snow slope, and the engine idled right back. There was no forward movement of the aircraft after touchdown — only backward movement down the ice slope toward the open sea. Fortunately, the pilot was able to call on his engine for assistance, and the aircraft was manoeuvred sideways 200 yards across the ice until it faced into wind and down a hill. The skis were retracted immediately the aircraft came to rest. Mawson was notified that the landing had been carried out without casualty, and a radio schedule arranged for 0315Z on the following day (16 August). (72)

The aircraft had to be tethered to ensure its continued stability. It was necessary to use hot water to melt the ice, into which the tie-down chocks would be frozen. With the aircraft tied down, a tent was pitched on nine inches of snow in a small valley 200 yards from the Beaver. The men doubted the ability of the tent to withstand the high winds for more than a few hours. The wind abated to 30 knots by the following morning, and a windbreak of snow blocks was constructed to ease the obvious pressure on the survival tent. Mawson was contacted, and the castaways waited the arrival of Wilson overhead on a flight to Davis. He had been advised of the ice bound airmen's needs, which included a larger, heavier duty tent and replacement cooking utensils. Wilson promised that he would return from Mawson two and a half hours later. In company with Alan Richardson, he landed near the downed Beaver; a severe downdraft on the final approach over the ice cliffs necessitated the use of maximum power at touchdown. Manning and McCleod returned to Mawson on the return flight, leaving Grove and Richardson to remedy the problem with the Beaver's engine.

Richardson, working gloveless in a temperature of –5 degrees Fahrenheit, had inspected the oil line to the engine by nightfall. After a cold night the two men, thankful that the wind had dropped even further, continued their task. Even though the conditions were relatively pleasant, it took five hours to start the aircraft's engine. Richardson, braced against the freezing slipstream, considered that he had found the source of the oil leak. No obvious fault was found once the engine was shut down, and a radio blackout prevented the duo from contacting Mawson.

The two men passed an uncomfortable night, Richardson ruing the day that he grew a beard. It was too cold to keep his head outside the sleeping bag, but when he covered his head, the ice in his beard melted, thus adding to his discomfort. (73) Wilson landed in deteriorating weather. The snow was blowing across the ice surface under an overcast sky as Grove and Richardson conferred. Richardson believed that there was sufficient oil for the engine to run for an hour. Grove decided to attempt to fly the aeroplane back to Mawson. Richardson joined Wilson, who had not left his aircraft, and the two aircraft returned to base. In Richardson's words 'luck was with Ivan, and he was able to make base with a cooked engine which wouldn't have lasted another five minutes'. (74) It took Richardson until 29 August to complete the engine change on A95-203. The loss of oil pressure was due to ice in the oil collector box, and the problem was cured by disconnecting the crankcase breather from the box.

Work commenced on the establishment of fuel depots in the Oygarden group on 12 September. This activity was followed by flights to the Taylor Glacier, and the survey of a sled route from Amundsen Bay to Mawson. Wilson discovered a group of nunataks which did not appear on any extant maps, and a depot site was chosen in their vicinity.

Antarctic Flight in action. Mr Knuckey, a surveyor, fixes a position in the Grove Nunataks after a Beaver exploratory flight. (RAAF Official)

The crew of a Russian IL2 from Mirny were welcome visitors, giving the airmen a diversion from the task of establishing depots at Beaver Lake,

Leckie Ranges and Prince Charles Mountains. A buffet lunch was provided for the visitors, and the hazards of polar flying forgotten for a period. Operations had been undertaken under extreme weather conditions with white outs, erratic winds, unreliable maps of the featureless terrain, and the difficulty in judging height on landing approaches providing a constant challenge. For example, Wilson attempted to land on light sastrugi 40 miles from Gillock Island, and found that the aircraft's skis were breaking through the ice. With great alacrity, he pushed some 6 ½ inches of boost into the engine, only just disengaging from the ice and becoming airborne.

Camp site at Amundsen Bay, 400 miles west of Mawson. (RAAF Official)

The advent of warmer weather in December brought the appearance of potholes in the vicinity of the hangar. Timber was employed to bridge the slushy area between it and the sea ice. Flying conditions still remained testing; on the 10th, Wilson was making a normal approach to land at Mawson when the aircraft was subject 'to severe turbulence when . . . a 1,500 feet per minute sustained climb was indicated for some minutes while attempting to descend with the throttle fully closed'. (75)

A request was received from the Belgian base at Breid Bay for assistance. An Auster, with four men aboard, was missing on a flight from Breid Bay to the Belgian inland base at Trillingane. The Antarctic Flight could not assist, as the area was beyond Beaver range. The request was passed to Mirny, with the offer of fuel. Bad weather prevented an IL2 from landing at Mawson until 12 December, and similar weather at Breid Bay prevented the Russian rescue flight from departing from Mawson until the day after. The crew, all from the Polar Aviation Group, and the Australians exchanged pleasantries during the wait. The Auster was discovered on the 14th, with its undercarriage broken. The crew, who abandoned the aircraft with very little food, and only sleeping bags available for warmth, were extremely lucky that the Russian aircraft found them. The news of the rescue arrived at Mawson on the 16th, with the IL2 staging back through the station two days later.

The Beaver, fitted with 25 gallon drums of diesel fuel on light series bomb carriers, is prepared to undertake a delivery. (RAAF Official)

On 2 February 1959, the Russian supply ship *OB* landed fuel to be used by aircraft staging through Mawson to the new Russian base at Lazarev. Of greater significance was the presence of *Thala Dan* in Horseshoe Bay on the following day. A95-202 was loaded aboard, and the ship sailed for home on the 14th. Wilson flew two local sorties at Davis on 17 February 1959, marking the final flight of the season. During the season, discussions took place which would have a significant impact on the subsequent Antarctic Flights. During July, the Antarctic Division raised the subject of operating the Beavers from the plateau above the Mawson base when the sea ice in Horseshoe Bay thawed. After an exchange of messages, the Director of the Antarctic Division advised on 19 November that:

I wish aircraft to experiment this summer with operations from airfield

> three miles south of Mawson. I would like aircraft to be flown onto plateau just before sea ice breaks up in harbour. I consider the value of flying through the summer months justifies the risks of leaving the aircraft in the open. [76]

A considered reply was drafted. Based on numerous plateau landings, it stated that:

> Experience gained during spring depoting operations has shown that Beaver aircraft constantly flown in overload conditions will not withstand the load imposed on it during frequent landing and take-offs from plateau ice . . . (and) to operate under conditions intimated is unwise and unnecessary . . . [77]

Plateau operations, it was explained, entailed such intense vibration in the aircraft that flight instruments could not be read during landing, taxiing and take-off.

Methods were suggested to overcome the known shortcomings of the Beaver. Considering that future aerial support would be required for inland astrofix and geological work, the suggestion was made that the provision of a large twin engine aircraft would solve many of the problems.

J.C. Sandercock, leader of the 1958–59 Antarctic Flight. (RAAF News)

The Director reluctantly agreed not to order plateau operations during the current season. But the seed had been sown, germinating in an attempt to undertake operations during the following year, and the provision of a Dakota aircraft to fly with the 1959–60 expedition. It may be argued that the efficiency of the Antarctic Flight was seen as proof that the aerial presence in the Australian Antarctic Territory could be expanded. The experience gained by Grove, and the rejection of his advice that plateau operations during the summer months was not practical, almost brought tragedy in the years ahead.

A95–202 poses against a backdrop of nunataks. Note the ski undercarriage, wide access doors and cameras mounted aft of access door. (RAAF Museum)

Whenever possible, realistic conditions were utilized to train the Antarctic Flight. For example on 25 August 1958, Dakota A65–121 landed at Cooma, New South Wales, with members of the Flight. The Beaver aircraft arrived during the following day. It was intended that a 2,200 feet strip which Peter Clemence had reconnoitred at Cabramurra be used for training with the Beaver fitted with skis. Unfortunately, the snow had cleared from the area, and bad weather prevented the aircraft being utilised — it was too warm for the snow to reform.

Despite this setback to the training programme, J.C. Sandercock led his men, G.A. Banfield, R. Rippon, H. McIntyre and S. Bell aboard *Thala Dan*, which weighed anchor bound for Davis and Mawson on Boxing Day, 1958. J.C. Kitchenside, who was to command the 1959–60 Flight, was also aboard as was Beaver A95–201.

Only 24 hours later, *Thala Dan* suffered the first of a series of incidents which were to plague the expedition when she suffered engine problems. The engines had to be shut down while the defect was corrected. While stationary, the ship was rolling through 40 degrees, and the Antarctic Flight lost 40 drums of Avgas overboard — fuel for some 100 hours flying. The fuel had to be abandoned, and the ship sought the lee of King Island where the remaining fuel containers were re-secured and spilt fuel sluiced overboard. The pack ice was sighted on 10 January 1959, and the ship berthed at Mirny on the following day. Members of the expedition were transported to the base by helicopter, giving Victor Perov, the Russian station commander, the opportunity of entertaining the Antarctic Flight. Russian hospitality was reciprocated aboard *Thala Dan* with scotch and sherry. (78) The ship sailed for Davis on 12 January.

Within sight of the station, *Thala Dan* ran aground on an uncharted rock, losing 60 tons of fuel oil. The vessel was trimmed to enable the 15 inch fracture to be caulked with concrete. By using an ice anchor, the ship was re-floated, and was the base for the Beaver's first flight on the 20th when mail was dropped to the incumbents at Davis. An ice reconnaissance was flown, and a flight was made to Lake Stinear to establish whether it would be suitable as a base for the Beaver while the ship was being unloaded at Davis. The Flight re-embarked its aircraft after deploying to Lake Stinear on 31 January, the day before *Thala Dan* commenced her run to Mawson.

The ship broke through into Horse Shoe Bay on 4 February. During the changeover period, one of the weazels broke through the ice and sank in 12 feet of water. Alan Richardson donned an immersion suit and linked a steel cable to the vehicle's towing hook, thus enabling it to be salvaged by another weazel. (79) Sandercock and Grove undertook the 36 mile round trip to the Masson Ranges to review possible landing sites, before the ship sailed on 14 February. A95-202 was backloaded to Australia.

The Flight's aircraft did not become airborne until 6 March. However, members contributed to the building of a new power house and were integrated into the camp routine. Bell became the base colour sergeant, Rippon the movie projectionist and rum master and McIntyre responsible for tractor maintenance — the 'fergy-fitter'. Sandercock developed into the local horticulturist, his plants thriving under the skylight in the mess.

Geoff Banfield and John Bechervaise, the expedition leader, were discussing the former's role as amenities officer when one of the most feared calamities of Antarctic life occurred — fire. Co-ordinated efforts by all expedition members failed to prevent the newly constructed power house from being gutted, thus dropping the electrical output of the base by 60 per cent.

The wheel-ski combination was fitted to the Beaver on 16 April, after the aircraft made a solo test flight fitted with floats. Four days later the aircraft transported a party to Taylor Glacier. On 9 May the initial direct flight was made to Davis, where inclement weather grounded the aircraft until the 22nd. During this enforced stay the radio aerials, which had been carried away in a blizzard, were replaced, a carburettor icing problem solved, and McIntyre discovered a partial solution to snow icing the wings by wiping them over with glycol. Beaver A95–203 was tied down on the plateau behind Mawson for four days between routine flights, no doubt as a trial for the deployment of the aircraft onto the plateau during the summer season. Due to the weathering, the engine had to be hand cranked to start.

A party was flown to the Foldoya Emperor Penguin rookery, 83 miles west of Mawson, on 8 June. The landing surface appeared smooth but, in reality, was a ridge of hard sastrugi filled with soft snow. After landing, it was found necessary to taxi for four miles to find a suitable take-off area.

This incident preceded a bout of bad luck, which commenced on 18 June when Geoff Banfield was eight miles from Mawson on a trip to Taylor Glacier. He felt alarming shivers in the airfoil as the port aileron counterweight fractured, forcing his return to Mawson. It was decided to replace the unserviceable part from the partially disassembled A95–203. Under Rippon's direction, the Flight members combined to complete the task under adverse conditions and limited facilities. Sandercock flew the aborted sortie on 8 August. In the meantime, McIntyre had a lucky escape when running the engine on A95–203 in the hangar. Snowdrifts built up by mid winter blizzards had made the hangar virtually airtight, and Sandercock found the airman prostrate underneath the aircraft — a victim of carbon monoxide poisoning. The spare floats and heavy wooden cradles were smashed against the wind fence during the blizzard of 27–30 July, which also damaged the workshop.

No contact had been made with the party at Taylor Glacier for four days, so Banfield flew Bechervaise to the site on 4 August to discover the reason. For the expedition leader, the last few minutes of the flight were:

> . . . the worst moments I have spent in the Antarctic. Geoff said in a strained voice, 'I can't see the station, the station's not there. The station is gone!'. We banked steeply. Long snow drifts ran down from a meteorological mast on a rocky outcrop. There was a black scar near the tide cracks. 'There's been a fire!' Geoff said. 'God, the place has been burnt down!' . . . No sign of life . . . We came down low for our second circuit, and I could see the other hut standing forlornly, deserted. The Beaver raced across the rocks and snow, suddenly so remote and grim . . .' We must get down, Geoff'. 'Look, they're there! They're alive!' came Mike's voice. [80]

The survivors, who had escaped with only the clothes in which they stood when the sleeping quarters caught fire, were flown back to Mawson. That the flight took place at all is a tribute to the dedication of the ground crew. Rippon spent hours removing snow from the interstices of the wings, and was forced to make a dash to the aircraft before it could take off. The pilot's view through the windscreen perspex panels had become obliterated by thick rime which could only be removed by using de-icing fluid.

When the generator motor broke down on the 12th, the Beaver flew to Davis to enable the suspect part to be repaired. During the flight, the phenomenon which Clemence and his men had encountered two years previously showed what Nature could do to a fragile aircraft:

> Between Mt Kjerka and the southern end of McKenzie Bay, a very strong, steady downdraft was experienced in which considerable height was lost even though climbing power and flap were used to try and maintain height. At the same time very excessive drifts were recorded (20 degrees and over), indicating that there appeared to be very strong outpouring of air in this region. These high drifts were noticed in later flights, although the downdrafts were never again experienced. (81)

Due to the vagaries of the weather, the week-long geological survey planned from the Davis base extended to four weeks. Having left Mawson on 23 August, the aircraft was not able to return until 19 September.

The weather and the accident to A95-201 had retarded the flying programme. However Bell and the station radio supervisor maintained the ground/air transmitters and receivers to such an extent that communications were the best yet enjoyed by the Flight. The hangar was utilised for the storage of spares and the overhaul of vehicles.

Spring did not bring an improvement in the weather. The Flight completed geological and survey flights in the Davis area before 18 days of continuous blizzards grounded the aircraft at both stations. For the crew at Mawson it was a frustrating period; each morning the drift was cleared from the front of the hangar, only to build up overnight, so the process had to be repeated. At Davis, the airmen assisted with the overhaul of the three 15 KVA generators.

Beaver A95-201 returned to Mawson on 23 September. Both aircraft were involved in supporting the scientific teams with their observations; it had been only possible to fly on eight days of the month. During one of these flights, as Sandercock and Bell flew fuel for the Beaver to a new depot, a sizeable oil leak was discovered. The 90 minute return flight to Mawson was made with Banfield in A95-203 standing by in case of a forced landing.

During October, the aircraft were busy over Casey Bay and Bryggeholmen Island. On 6 October Sandercock had difficulty taxiing at

Casey Bay due to a combination of slippery ice and wind. On the 21st, the idling system in A95–203 was found to be affected by the ice and was remedied before both aircraft took off. One aircraft suffered a complete radio failure, and returned to Mawson.

Joys of Antarctic flying. A Beaver is covered by drift snow at Mawson. (RAAF Museum)

White out conditions forced both aircraft to land near Crooked Island, 120 miles from Mawson, on 27 September. The aircraft were secured to deadmen — lengths of timber imbedded into the ice — and survival tents erected. The crew remained at the island until conditions improved on the morning of the 29th, living on army rations for the period of their grounding.

Visitors, in the form of an IL2 crew from Mirny, landed in October and stayed overnight. The Russians continued their flight to Lazarev, only to be forced back by bad weather. The aircraft was scheduled for another task so it returned to Mirny. Doctor Draklin, the commander of the Russian expedition, and two of the crew stayed at Mawson where they accompanied Sandercock on a flight to the Auster Rookery and on depot flights. Also during the month, a major service was undertaken on A95–201, and work commenced on the plateau airstrip. Caravans were re-secured at the Gwamm site (as it had been named), and deadmen laid for the Freighter caravan which was to be used as a workshop and radio shack.

During November the Flight's bad luck returned. When landing at

Beaver Lake on the 3rd, A95-203 was caught in a crosswind gust of wind and damaged the starboard wing tip. The cargo was unloaded before it was discovered that the starboard aileron was fouled. This was freed, but the subsequent repairs at Mawson took Rippon until the 10 November to complete. At this time, it was decided to convert A95-201 into its photographic configuration to conduct an astro survey in Kemp and Enderby Lands, using a camp on King Edward VIII Gulf as a base. This task was completed as the other Beaver transported scientists from Mawson to undertake geological sampling and photographic sorties.

On the 22nd, a faulty polar compass in A95-203 and inadequate maps resulted in the crew of the aircraft being uncertain of their position. After a positive position was plotted, the aircraft was found to be well north of the required track and returned to the depot with tanks almost empty. Banfield and the other Beaver was kept on search and rescue standby at Mawson. Sandercock was also involved in another incident when taxiing out for a flight to Davis. The tailskid caught in one of the tie down loops in the sea ice, and caused damage to the rear of the aircraft. With only one aircraft serviceable, it was considered prudent to restrict flying to within a 50 mile radius of Mawson.

Beaver A95-201 was flown to the plateau airstrip in readiness for the summer operations on 5 December. Both aircraft were involved in routine flying during December. At Davis, on one of these flights, A95-201's tail strut collapsed. The remaining Beaver delivered the requisite spare on the 14th, and undertook the photography which was to have been undertaken by A95-201. After being repaired, A95-201 returned to Mawson on the 17th, flying through white out conditions and landing just before a light snow fall fell at the station.

The time before the festive season was occupied establishing fuel depots at Beaver Lake, obtaining astrofixes and preparing for New Year operations. During the morning of 28 December, Sandercock, McIntyre and Bell drove a weazel to the ice strip on the plateau above Mawson with the intention of undertaking routine maintenance on the two Beavers tethered there. The weazel broke through an ice encrusted meltwater channel en route, delaying the party's arrival until 11.30am. By this time the moderate wind which was blowing when they departed from Mawson had increased to a velocity of 80mph, and the aircraft were in danger. A95-201 had snapped its three ton tie wires and the wing tip fasteners had broken loose. Sandercock climbed into the glissading aircraft, started the engine, and turned the Beaver's nose into the prevailing wind — an action which would place the pilot in danger of serious injury if the aircraft could not be tethered securely.

Prior to Bell driving the weazel to Mawson for assistance, the increasing strength of the hurricane tore the heavy railway sleepers to which A95-203 was anchored out of the ice. These deadmen were buried to a depth of two feet. McIntyre and Bell charged after the aircraft in a D4 tractor and succeeded in dropping the dozer blade onto one of the anchors. Seven and a half ton breaking strain steel cables and wing tip tie ropes were used to further secure the aircraft.

Sandercock's aircraft was intermittently airborne above the steep icy slope which led to the coastal ice cliffs when Bell returned from Mawson with Bechervaise and Rippon. The three men, armed with heavy iron stakes, cables and mauls, drove the almost uncontrollable weazel as close to the spinning propeller of the Beaver as possible. The cramponed men, leaning against the hundred mile an hour wind, drove stakes into the ice, and secured the Beaver. Sandercock, after two hours of 'flying', could now be relieved.

Another man and tractor arrived from Mawson, and the seven men battled the wind to the haven of the on-site caravan. Providentially, its iron runners had frozen into the ice, as it was unlikely that the five ton steel guys employed to secure the caravan would have been able to solely absorb the stress placed on the van over the next 20 hours. The party had no sooner reached the comparative safety of the caravan when the wind gust velocity increased to 120mph. Its direction was erratic. All the previous effort to save the aircraft was to no avail. Wing lift spoilers proved ineffective; at times both aircraft dragged the six ton tractors to which they were anchored over the ice. The unprecedented buffeting of the hurricane overcame the cables which secured A95-203. The Beaver slithered downhill, bending both mainplanes at grotesque angles, before being re-secured after herculean efforts. The wind was now charged with ice particles being blown off nearby Mount Henderson. Despite the men's efforts, A95-203 appeared intent on self-destruction. Bechervaise later reported that:

> (A95-201,) directly upwind of the caravan, started to break up, one wing collapsed in the middle; the other was wrenched off and hurled through the air for a hundred yards without touching the ice. It missed the port side of our shelter by five yards. A wheel skid charged past to starboard. The aircraft lifted itself madly in the air, even on half a wing, and dashed itself repeatedly in an intermittent spray of ice fragments and petrol, destroying the undercarriage and tailplane. The propeller slowly revolved against the engine compression. We feared fire, for the battery of 201 could not be removed. [82]

For a further 17 hours, the men kept watches in case the caravan should break loose. Men snatched what sleep they could clad in their windproof clothing and cramponed feet before the weather improved and it was possible to drive the weazel back to Mawson.

Result of the summer cyclone of December. The Beavers were definitely not airworthy. (RAAF Official)

The loss of the aircraft was a serious setback for the expedition. The Beaver's crew had established supply bases for an intense geological and photo mapping survey in the Prince Charles Mountains. Bechervaise, the expedition leader, was sanguine at the prospects of future Flights being able to operate from the plateau. He argued that the aircraft had survived 110mph winds while on the sea ice, and that the lessons learnt from the disaster would be applied in the future. Time was to prove that his aspirations could not be fulfilled.

The disappointed Flight salvaged the remains of the aircraft, repaired the damaged hangar doors, and packed stores which would be returned to Australia. For Sandercock it was not a happy period; he contracted hepatitis, and was hospitalised at 6 RAAF Hospital, Laverton, on his return.

Magga Dan was farewelled from Melbourne by the Minister for External Affairs, Mr R.G. Casey, on 7 January 1959. The expedition was to undertake a threefold task – to resupply and assume control of the American base at Wilkes, overhaul the automatic weather station at Lewis

Islet and attempt to land on, and explore, the coast of Oates Land. The RAAF supplied two Antarctic Flight veterans, D.W. Leckie and Auster A11-201, to assist with these tasks. Leckie was to obtain hand held photographs of the Oates Land coast. (83) To maintain the aircraft, N.W. Meredeth and J.S. Williams were selected. The *Magga Dan* pierced the pack ice, in foggy but calm weather, after five unpleasant days at sea prior to anchoring off Lewis Islet on 13 January. Unloading operations commenced during the following morning. It was discovered that considerable work was required to refurbish the station; two masts had collapsed, and technical work was required on the equipment. While this work was in progress, the scientists took the opportunity to undertake short field trips, but the rough sea state and the low cloud made it impossible to fly the Auster.

The captain was forced to weigh anchor and move the ship due to a 40 knot wind. With difficulty, the men had to be evacuated from the islet.

Magga Dan's next port of call was Durmont d'Urville, where the ship berthed on the 18th after weathering a gale for two days. The Director of the Expeditions Polaires Francaises, Paul Emile Victor, commanded the base. The Frenchmen and Australians enjoyed an exchange of personnel, when 12 of each nationality were entertained by an equal number of the other. During the sojourn with the French, Phillip Law, the Australian leader, flew by helicopter onto the continental shelf behind the station to examine the feasibility of establishing an ice airstrip. The most suitable area could not be sited into the prevailing wind, and although the surface was suitable for operations by a Dakota, flying would only be practical on rare calm days.

The Australian expedition returned to Lewis Islet, where it arrived in a 32 knot gale on 19 January. Despite the weather, four men were landed to complete the refurbishment of the meteorological facility. A 60 knot wind on the following day forced their return to the ship. *Magga Dan*, having been forced from her anchorage, found it difficult to return. Finally it did so, to enable adjustments to be made to the equipment on the islet. Its task completed, the vessel sailed for Wilkes, breaking through the pack ice to the base on 24 January.

The initial task for Leckie and his men was to land the Auster. Aviation fuel was unloaded, and the aircraft's floats removed. Law describes the landing procedure:

> We finally chose a long strip of fast ice attached to the coast south of Wilkes behind some islands. It was approachable by pontoon and provided a good airfield for the aeroplane on wheels. The condition of the ice at the edge was somewhat dubious but we risked it and

managed to get the aircraft safely ashore. Leckie then flew it up to a better strip on the plateau behind the station, after first fitting skis. From the airstrip behind the station he was able to taxi down to peg it out near the radio aerials. With the aircraft on skis ashore a lot of flying immediately became possible. (84)

A11-201, the Antarctic veteran, is brought ashore at the American base at Wilkes, 24 January 1959. (RAAF Official)

Leckie flew expedition members to inspect landing sites on the plateau next day. On one of the later flights, a ski hit an obstruction on take-off. The aeroplane bounced back, bending the starboard strut. American mechanics fabricated and welded a repaired section, which meant that the aircraft was available again on the following day. On the 29th the Auster flew from the coast to a station situated at an altitude of 4,000 feet. The landing was bumpy on the firm snow, but did not dissuade other members of the expedition from making the same flight later in the day. Leckie also flew some photographic sorties for the Americans.

For everyone except Leckie, Sunday, 1 February was the first day of rest since the vessel departed from Melbourne. Leckie flew the Auster to the Vanderford Glacier, to enable a survey to be made to examine the glacier's movements. The U.S.S. *Staten* arrived on the 2nd, and the physical handover of the station to the Australians was formalised on the 3rd.

On 5 February the aircraft was returned to the ship. Law takes up the story:

> Leckie had planned to slide the aircraft on skis from an ice edge, near where the aeroplane was tied down, onto the rear of an Army DUKW, across which timbers had been laid. The main problem was whether the ice edge would support the weight of the plane. Already portions

> of the ice edge had fallen in and much of it was undercut by the tide. I made a very close examination of it and finally ran a weazel several times to the front of the cliff to test whether it would take the weight. We then brought the aircraft down and after some difficulty slid it onto the DUKW. Twenty minutes later an extensive block of this ice front up to within several feet of the place we had used fell into the sea. (85)

Law then returned to the plateau in search of possible Dakota landing grounds. The Auster was converted to floats, and put to bed on Number Three hatch. This was the final action with the flight of Williams. The party who were to winter over at Wilkes were short of fitters, so Law made arrangements for the airmen to remain with the party and give much needed technical expertise.

Wilkes Base from the air. This base was taken over by Australia in February 1959 and is now known as Casey. (RAAF Official)

Magga Dan departed for Lewis Islet, beating a path through the pack ice before arriving there on the 9th. Wooden aerials were erected on the islet, before the vessel voyaged in the direction of Oates Land, its progress was hampered by the worsening pack ice. On 11 February, the presence of a small open pool gave the opportunity to fly the Auster on an hour long ice reconnaissance. Although open water was sighted beside a giant iceberg 40 miles from the ship, the lead did not give access to the coast. Another flight to discover other leads was not possible due to the low cloud level and occasional snowstorms.

During the following morning the ship commenced its battle with the ice and other elements which continued until the 20th, when the wind and tide eased the pressure of the ice. The Auster was hoisted overboard into an open pool. Law describes the subsequent flight:

> We flew south-westerly to reach the coast near the most westerly of the mountains. Turning east we flew a course of 115 degrees at an altitude of 2,000 feet and, using a hand held aerial camera, I shot some 60 oblique photographs . . . We ran down to the end of the main features and Leckie then said he would have to return because of the fuel supply . . . We headed . . . back to where we thought the ship would be. Altogether this had been a fairly nerve-racking flight. Our first moments of disquiet had come when we were taking off. We were considerably overweight . . . and we barely managed to scramble off the water and limp into the air before passing over the floating bits of brash ice near the end of the pool. Now, when we came back we could not find the ship and the radio compass was not giving any indication. The whole configuration of the pack had altered during the time we were away. Down beneath us there was a mosaic of confetti-like ice floes which we scanned desperately looking for the ship. Fortunately we were in touch with the ship by radio. We asked for their position in relation to nearby icebergs but they could give us nothing over the radio which would help us in relation to what we could see. We then tried obtaining a bearing from them on the western edge of the mountains but that also did not seem to help much. I then asked them to send as many men as possible onto the bridge to look out for us with binoculars and we flew over the edge of the ice pack and told them we were there. Then they picked us up and talked us back to the ship. It was apparent to us if we landed among the ice floes the ship would have little chance of finding us, for when they discovered us it was apparent that we had been looking in the wrong place, at least ten miles east of the ship's position. (86)

The ship then moved westward, and the members of the expedition made their first landing on Oates Land.

Leckie flew the ski equipped Auster, with Law again his passenger, on 21 February. At 2,600 feet they passed the most westerly of the mountains, along a grim coast to where it commenced to trend northward. Having lost radio contact with the ship, both men felt some trepidation. At 7.50am, a rectangular shaped rock outcrop, and other nunataks, were sighted. Law wished to continue to the northern extremity of the headland over which they were flying, but Leckie, with the engine sounding 'funny', headed back for *Magga Dan*, 76 statute miles distant. The engine malfunction was due to the failure of one of the fuel pumps. Leckie considered that the lack of power was due to carburettor icing, and switched fuel tanks. By so doing, he brought another fuel pump into play. The Auster was nursed back to the ship, being followed by the weather, which was closing in from the east.

Law decided to undertake two more flights on the 22nd, but the rising south-east wind, low overcast, dropping visibility and barometer prevented any flying. The Auster was not destined to fly over the Antarctic again. At 8am the ship was stopped and the aircraft fitted with floats and lashed down for the voyage to Macquarie Island.

The voyage commenced in a 30 knot gale. Providentially, the gale had abated by the 24th, when a broken turbo blower on the main engine caused the *Magga Dan* to heave to until the damage could be repaired. The ship reached Macquarie Island on the 27th. It departed at 4pm, bound for Melbourne where *Magga Dan* berthed on 5 March.

For Doug Leckie, it was to be his last visit to the Antarctic as a member of the RAAF. In his book *Antarctic Odyssey*, Law published the following tribute:

> Leckie was a born pilot. He was superb in landing small planes in difficult or restricted areas, but then he had a lot of experience, first in jungle flying in New Guinea during the war and, later, as a pilot with the Snowy Mountains Authority, making landings on emergency airstrips in the Australian Alps. I was to fly with him on numerous occasions during the years . . . and he and I survived a number of hair-raising episodes. (87)

CHAPTER SIX

Summer cyclone

THE 1959–60 Antarctic Flight was the most ambitious to date, both in manpower and equipment. The Flight consisted of three pilots, a navigator, a signaller and a maintenance crew of seven for the Dakota (A65–81) with which it was equipped. The members of the flight were J.C. Kitchenside, H.V. Carne, E.C. Bloomfield, N.W. Hanson, K. Assender, G. Dyke, H.B. Harris, R.D. Murphy, J.T. Arthur, B.P. Rutter, D.V. Monks and K. Felton. The Dakota had been christened 'Ann Cherie', after Kitchenside's two daughters, and had been modified at De Havillands where it had been equipped with a ski undercarriage, cabin heating, a Doppler navigation system, cameras and the provision of JATO equipment.

The unit was officially raised on 15 June 1959. In addition to normal training, the Flight aircrew trained under icy conditions, and were present at the Aircraft Research and Development Unit, Laverton, when JATO trials with the aircraft were undertaken.

'Ann Cherie' was loaded aboard the *Thala Dan* on 10 January, 1960. It had its wings previously removed by staff of No.1 Aircraft Depot, and fitted snugly on top of Number Two and Three holds. Here it remained for the 15 days of the passage to Mawson, while the airmen played board games, read, played cards and underwent essential boat drills. An informal dining-in night was one of the activities in which the servicemen were involved.

High winds and seas forced *Thala Dan* to heave to outside Horse Shoe Bay on 24 January, but cleared the bay of ice, and the ship was able to berth during the following day. It was not until the 29th that 'Ann Cherie' was brought ashore to the western ice shelf of the bay. The delay was caused by the necessity to construct a 50 by 20 foot platform on four pontoons, and the preparation of a tie-down point. The aircraft was hauled ashore at high tide, along a specially constructed road, to her new berth. The whole process took six hours.

1959–60 Antarctic Flight members pose with A65–81. L-r: Felton, Rutter, Arthur, Kitchenside, Bloomfield, Dyke, Harris, Carne, Murphy, Monks, Assender and Hansen. (Kevin Felton)

The Dakota being loaded at Melbourne. (Graham Dyke)

During the unloading of general cargo next day, a mishap occurred which may have been disastrous for the Flight. The crane lifting the port mainplane gave way, allowing the wing to drop some distance onto the ice. Fears were held for the wing's structural integrity.

. . . and being unloaded at Mawson. (Graham Dyke)

The Director of the Antarctic Division, P.G. Law, and his pilot, Richard Cresswell, arrived in Beaver VH-PGL on 1 February. This aircraft was taken on charge by the Flight as A95-202, and was a replacement for the two aircraft which had been destroyed during the previous December's hurricane. During the next day Kitchenside, Bloomfield and Graham Dyke flew the float-fitted Beaver to the Masson Ranges, to select the site for the Dakota base. Other members of the flight involved themselves with the changeover operations. Felton fractured a toe when a tractor tow bar dropped on his foot. (88)

Law, Cresswell and B. Stinear were flown to where the *Magga Dan* was standing, 60 miles off shore, on the 5th. It had initially been planned that the trio would be returned to Davis, but the weather at that station prevented the return flight. By the 9th, 800 tons of stores and equipment had been unloaded and *Thala Dan* departed from Mawson on 16 February. Before she did so a Russian IL2, en route to the base at Lazarev, landed at Gwamm to refuel.

The plateau airstrip, named Rumdoodle, had to be prepared. Sited 15 miles south of Mawson, on a rock area half way along the Masson Ranges, it was the destination of seven men on 22 February. The party was to establish the accommodation caravan, service generators and note meteorological conditions which would affect aerial operations. The rotation of members between Rumdoodle and Mawson was made difficult by the heavily crevassed terrain and common white out conditions. The first blizzard of the season was a sobering experience for the men camped at the plateau airstrip.

The first Rumdoodlers. L-r: Felton, Arthur, Kitchenside and Stretton (one of the expedition members) survey the proposed landing area. (Kevin Felton)

Work on preparing 'Ann Cherie' for flight commenced on 19 March. A snow ramp had been constructed to enable the wings to be set at the required height for fitting them to the fuselage. Luckily, the port wing had not been damaged, and the tedious job of fitting the wings commenced on the 25th. Felton recalls that the:

> Placing (of) some 300 bolts ¼ inch in diameter and 1 inch long through holes, whilst wearing gloves or mittens, in winds gusting to 50 mph and a temperature around 0 degrees Fahrenheit, is virtually impossible. There was only one thing to do, 'Off Gloves' and work in relays. Four men inside the aircraft, warming hands on the auxiliary heater system, whilst four outside slid bolts in and nuts on, until the fingers went numb with cold. (89)

The matter was made more comfortable by using the Airco heater to supply heat to the work area. Both mainplanes were fitted by the afternoon of the 27th. Drifting snow prevented any further work on the aircraft until 4 April, when the drift was excavated. The engines were run on the 11th and the aircraft was declared serviceable four days later.

Members of the 1959–60 Antarctic Flight en route to Mawson. L-r: Hanson, Felton, Rutter, Monks, Arthur and Carne. (Graham Dyke)

Bert Carne and Felton had spent many hours tracing minor problems with the aircraft. One was a consistent low pressure reading on the cockpit instruments, which was caused by the oil in the lines congealing. It was found necessary to remove the ailerons, which required ice and snow to be thawed from their interiors; the spare set was found to have collapsed fabric, and repair was difficult in the prevailing weather conditions. On 12 May, the aircraft was ready to be moved to the sea ice prior to being flown to Rumdoodle. To undertake the task:

> . . . the aircraft was winched by a D4 tractor away from the tie-down and started down the hill to the sea ice where it was intended to take-off. After being winched approx. 80 yards the Dakota was secured on compact snow and the winch cable released to allow the D4 to move forward — before this could be done the aircraft started to move forward and the brakes failed to function. As the D4 had been stationary winching the aircraft, it could not get out of the way. Corrective action

> was taken on the tail steering arm and the aircraft turned about 15–20 degrees before it hit the D4. The point of impact was on the starboard landing light, number one stringer, both top and bottom torn and some skin torn. (90)

The aircraft was not flyable. It was taxied to within 70 yards of the hangar, and tied down — a task which took four hours. Repairs commenced two days later, and were hampered by having to hand rivet the repair. Before the aircraft was test flown on 14 June, the cockpit had to be dried out, having filled with hydraulic oil from a fractured union. While the work on the Dakota proceeded, a wind fence had been erected at Rumdoodle. The presence of the Beaver made the trip to the airfield one of seven minutes, compared to the potential hours by surface transport.

The ski equipped Beaver was available for the support of field parties. Kevin Felton flew to Kloa, on King Edward VIII Gulf, to join two expedition members, Doug Machin and Hank Geyser. Before the party could be flown back to Mawson, it was necessary to tranquillise the dogs. One evaded capture for some time, leading the men a merry chase among the icebergs and creaking glacier, before being claimed in a rugby tackle by the intrepid Felton. (91) The party was withdrawn on 23 April.

Members relax in quarters at Mawson. L-r: Bloomfield, Kitchenside, Harris and Carne. (Graham Dyke)

Blizzard bound days made exterior activity nigh on impossible. The Flight members sorted spares in the relative comfort of the hangar. During the winter period, a Saturday night respite was instituted at which the dreaded Rumdoodle Punch made an appearance. This potion is described as containing:

> Four bottles of sherry, one pint of medicinal alcohol, one bottle of fruit cordial, two tins strawberries, one tin fruit salad, four pints of water. Mix together and stand clear! (92)

'Ann Cherie' flew on 2 July, but there were problems with her undercarriage, which would not retract.

Kitchenside flew the Beaver to the Stanton Group of islands on 4 July to pick up a party which had been man-hauling a sledge to Taylor Glacier. After these flights, Assender was dropped at Rumdoodle, where the tail wheel skid, wheel and ski of the Beaver broke, jamming the rudder and elevator controls. The damage was repairable.

The Flight was heartened by the test flight of the Dakota by Graham Dyke on 9 July. From an altitude of 16,000 feet, the view was magnificent. The aircraft remained temperamental. A flight proposed for the 16th had to be abandoned, due to the brake drum and shoes freezing together. Heat had to be applied to areas such as the carburettors and cylinders; it was necessary to place immersion heaters in the oil.

The Beaver was flown to an area north-east of Cape Boothby to enable a landing site to be selected for the withdrawal of a field party by the Dakota on 2 August. Two days later, Beaver Lake was visited for the same purpose. The Dakota flight to Cape Boothby was made on the 7th, but take-off was delayed by the aircraft having a flat tyre, which had to be inflated by hand. JATO was used for the take-off from Cape Boothby, and the aircraft landed back at Mawson after sunset, using a temporary flare path.

Both aircraft landed at Husky Dome, near Beaver Lake, on the 14th. The landing area was not considered as being suitable for a fully loaded Dakota. Felton, in the Beaver, describes the grandeur of the scene on the return trip:

> Flying at 10,000 feet, we were clearing mountains by 500 to 1,000 feet, and seeing some of the most spectacular sights I have yet experienced. The Prince Charles Mountains run east-west about 200 miles south of Mawson, then extend in a southerly direction for about 200 miles, then run east-west again. (93)

The Dakota flew a south-west reconnaissance mission on the 16th. Penetrating 425 miles from Mawson, it overflew the plateau, which had an ice thickness of 9,500 feet; seven hours of eye strain only found perpetual ice and snow.

On the 17th the Beaver was flown to reconnoitre a path for the southern tractor journey which was scheduled for later in the season. Sixty miles inland the flight had to be aborted due to heavy ground drift which obscured the terrain below the aircraft. The return was made through the Masson/David corridor. Graham Dyke and 'Baz' Rutter flew to Beaver Lake on the 20th, finding a Dakota strip on the northern side of Mount Merideth, and this flight was followed by others to Taylor Glacier.

The Dakota was flown to Davis, which was under the control of Ian Douglas, the son of Eric, on the 26th. For the crew — Kitchenside, Bloomfield, Dyke, Hanson, Felton, Harris and Arthur — the environment at Davis was pleasant after the rigours of Mawson. They tied the aircraft down on the sea ice and ran a cable from the power house for the immersion heaters, before embarking six drums of fuel on the aircraft. It was intended to transport these to the airstrip at Mount Merideth. This flight could not be undertaken until 31 August due to the deteriorating weather. The delay gave the ground crew an opportunity to check the port generator and left magneto. The flight was without mechanical breakdown, although moderate turbulence was met on both approach and take-off. The return to Davis was marred by icing in cloud over Prydz Bay, and the aircraft landed with only fuel for 30 minutes flying left in her tanks.

A combination of weather and the now common unserviceability of 'Ann Cherie' prevented all but one flight during the first 20 days of September. Even this flight had to be curtailed due to the rough running of the starboard engine. Graham Dyke could not maintain height on the single engine. The recalcitrant engine was restarted, and the aircraft nursed back to Davis. It was considered that a spark plug change may remedy the problem, so Bert Carne flew as a passenger with Assender in the Beaver, to transport the spares from Mawson. The new plugs did not solve the problem; neither did a carburettor change. The final diagnosis was that the number one cylinder was unserviceable. The Beaver was therefore flown back to Mawson to return with a replacement cylinder, piston and valves. Felton noted an unusual phenomenon during the flight:

> During the trip the sun, aircraft, and cloud were all in perfect position to form a type of solar halo. Below us, on the cloud, was a circular rainbow and in its centre, the shadow of the Beaver. The effect of the halo was in sight for more than an hour . . . (94)

The Dakota was ready for work on 21 September.

A flight was made to Mount Merideth by the Dakota on the following day, with Leon Fox, one of the Davis incumbents then flown on to Mawson, as he required dental treatment. On the subsequent return flight, another six drums of fuel were depoted at Mount Merideth. On the 25th, the aircraft landed at the depot, where the temperature was -14 degrees Fahrenheit and the wind 50 knots, to refuel before another penetration flight south of the Prince Charles Mountains. Two hundred miles out, the port engine gave trouble, and the Beaver was placed on standby in case it would be required for a search and rescue role. Overflying mountainous and crevassed terrain, Graham Dyke coaxed the aircraft back to Davis, where faulty spark plugs were replaced.

The Beaver, based at Mawson, was flown to check the HF/DF equipment. A visit was made to Rumdoodle to enable the wind fence, damaged during a blizzard which had racked the area for the first four days of September, to be repaired. The prevailing high winds made this task tedious, and the incident should have been seen as a warning of the possible repetition of the fate of the Beavers during the previous December.

An IL2 from Mirny, en route for Lazarev, landed at Mawson on 25 September. It departed for Syowa on the 27th, but suffered engine failure over King Edward VIII Gulf. Forced to land on the sea ice, the crew had to jettison two tanks of fuel to enable the aircraft to take-off from the rough surface. The aeroplane returned to Mawson, where permanent repairs were undertaken, before continuing on its interrupted way on the 29th.

Both the Australian aircraft were based at Mawson early in October. An engine run on the Dakota resulted in the feathering motor being burnt out; a Beaver flight to Davis was necessary for a replacement. The Dakota was test flown on the 6th, and then grounded, as it was due for a major service. On completion 'Ann Cherie' was employed on photographic duties. One, over Enderby Land, was aborted due to cloud. On the 22nd, it landed at Camp 22. Using JATO to take-off, it was subsequently flown to Davis to complete the flying programme there. On the 26th, the starboard engine lost power. The aircraft staggered back to Davis, where the engine was repaired. Although no further ill effects manifested themselves, it was considered prudent to return the aircraft to Mawson and change the suspect engine. Blizzard conditions prevented the Dakota's return until 1 November. The Dakota's engine change was completed, and the aircraft test flown on 13 November. The spare engine had not been modified to Air Force standards, and this delayed the work.

For the Beaver based at Mawson, it was a busy time. Kitchenside flew to Camp 22 on 4 October. In company with Assender and Murphy, he refuelled the Beaver at Mount Merideth then proceed to an area 320 miles south of Mawson, to select landing sites. Flights were then made to the Taylor rookery and Rumdoodle, before an attempt was made to fly to the field party at Camp 22. Closing weather conditions forced the aircraft to return to Mawson. On 18 October, zero visibility at Camp 22 made it impractical for the aircraft to fly there, so the Beaver was utilized over Amundsen Bay in seeking a camp site for use during later summer operations. Two more attempts to fly to Camp 22 were prevented by white outs and inclement weather conditions.

On 2 November, a weazel, which was being operated by a field party, was travelling on the sea ice en route to Auster Rookery when it broke through the ice. No radio communication having been made with the party, the Beaver was flown to make contact. Jennings, the engineer, remained behind to assist, but the weazel sank through the thin ice and was lost. The Beaver flew the salvaged gear back to Mawson. On the 8th three men were flown to the site of the mishap, and more equipment was later airlifted back to base. Maintenance flights were made to Rumdoodle, and on the fourth of these the bolt retaining the tail wheel pick-up arm to the fuselage sheared, necessitating a hangar inspection of the Beaver.

Kitchenside, Felton and Dyer flew in the Beaver to Amundsen Bay on 15 November, to give support to Kirby, the surveyor. Fuel was flown in by the Dakota, but flying was not possible for the first three days. The men studied the fauna, fished 'eskimo style' through the ice (catching up to 100 fish in two hours) and enjoyed the 40 degree Fahrenheit temperature. An attempt to undertake astrofixes on the 19th was initially marred by bad weather. On the return to camp, an Emperor Penguin rookery and a glacier, located on the east side of Risen Larsen, were noted. Astrofix flights were undertaken until the 23rd, when the fuel situation prevented further flights. As a diversion, Felton had walked to one of the Emperor Penguin rookeries, where he found:

> One little chick, (which) couldn't stand up and, as he was being shunned by the adults, I carried him — or her — back to our camp. 'Perce', as we named it, had a feed of sardines and milk, feeding very hungrily. (95)

Sadly, the young one did not survive the night.

The Dakota delivered two drums of fuel on the 26th, and the Beaver flew to Mount Christensen where severe crevassing prevented a landing. The aircraft diverted to the Nye Mountains. Returning to the base camp, the party refuelled and then spent the night in the Napier Ranges, at an altitude of 5,400 feet above sea level. An hour after take-off on the subsequent return flight to Mawson, the Beaver entered a cloud bank which forced Kitchenside to climb to 10,500 feet. Ice and snow formed on the wings, tail and windscreen. Another hour passed before straining eyes spotted the Robert and Wilma Glaciers through a break in the cloud. The aircraft was still iced up and sluggish. Kitchenside was forced to juggle the controls and fuel to maximise the Beaver's range. The aircraft landed at Mawson with four gallons of fuel in its tanks.

'Perce'. (Kevin Felton)

The Dakota had been employed photographing the Grove nunataks, and final preparations were being made for the transfer of full flying operations to the Rumdoodle site. Mr Korothevitch, the officer-in-charge of the Mirny base, and a party visited Mawson on the 27th. After staying for two days, their departure coincided with another series of photographic sorties being flown by 'Ann Cherie'. On the 30th, both aircraft flew to Camp 22. Repairs had to be undertaken on the port engine of the large aircraft before she flew in fuel, spares and personnel to the area. The Beaver remained at the site to give support to the geologist and surveyor, making flights over the Goodspeed nunataks, Mount Menzies, Mount Rubin and the Mawson Escarpment.

The sea ice was becoming too thin to support Dakota operations so the flight transferred to Rumdoodle on 2 December. From here, the Dakota flew to Camp 26 to return personnel, dogs and equipment to Mawson. The Beaver remained at Mawson until 8 December, when the wind fence at Rumdoodle was completed. The flight was consolidated, and prepared for summer operations. But Mother Nature was to play her hand.

The weather during the evening of 8 December was overcast, with reduced visibility and falling snow. The wind was gusting in the vicinity of 20–25 knots. Although meteorological reports predicted a blizzard for late in the afternoon, and further evidence of such a weather change was evident, the decision was made not to post a night watch with the aircraft. Factors taken into consideration were that no suitable transport was available, and that the radio link between the aircraft and the camp was inoperative. The Flight retired to the camp at 4.30pm. At 2.30am on the 9th, the wind gusts were shaking the caravans to such an extent that several vans had to be re-guyed. By 6.00am gale force winds were lashing the camp; visibility had reduced to a few yards, and the strength of the wind made outside movement extremely hazardous. The wind could be heard howling like an eerie banshee over the mountain tops.

Felton, with Graham Dyke, Kitchenside and Murphy clinging to the towed sled, ventured to check the aircraft at 8.40am. The party battled 100 knot winds to reach the wind fence, to find the Beaver on its back, with the wings ripped off. The wind was so violent that the men were blown bodily across the ice as they attempted to remove the aircraft's battery and re-secure the wings to the wind fence. There was no sign of the Dakota.

The aircraft had been parked into the wind, but nothing could be done to find it until the blizzard eased. Returning to camp, the party joined the rest of the Flight to wait out one of the most fearful experiences any had experienced. Felton, knowing that the guy ropes on the caravan in which he sheltered were slack, stumbled to the power house for a tool to tension them. One second he was gripping the power house with all his strength, the next blown bodily a distance of 15 feet onto a large rock. He crawled and stumbled against the ferocious wind back to safety.

Visibility had increased by noon, but the wind velocity remained constant. Another attempt was made to find the Dakota. The Beaver had broken up, a tangled mass of broken metal. The workshop caravan had snapped its guy wires and was completely missing. Following the skid marks, the tractor party proceeded until bad visibility forced them to return to camp. Running into the blizzard, the force of the wind threatened to stall the tractor.

Radio communication with Mawson was severed at 3pm, as all the radio aerials had been blown down. Monks rigged a jury radio mast, completing his task at 7pm, but contact was still not made. All the Flight members faced the situation with as much composure as possible, and Felton describes the scene which confronted them:

> The rocks on which we are camped are littered with packing cases, smashed to pieces and 44 gallon drums, some empty, some full. At about 4 pm . . . 'Baz' Rutter looked out of the window of our van and called

our notice to the extent of the wind. Drums of diesel fuel and petrol weighing 350 pounds were moving at a fast walking pace along the ice standing on their ends. The caravan nearest us shifted four feet sideways in one gust, stretching steel guy wires like rope. (96)

The men spent a sleepless night, the terror outside a reality which could not be ignored. By 1am on the 10th, the wind had eased, and snow commenced to fall. Five hours later, the camp was sheathed; the drift had cut the power supply. The drift was cleared, and contact made with the outside world at 10.15am. The camp had been under blizzard conditions for 42 hours.

Wreckage at Rumdoodle — the remains of the Beaver in front of the intact wind fence. (Graham Dyke)

At 7pm, the weather had cleared sufficiently for the tractor to be serviced, and an hour later another search was made for the Dakota. The search party discovered the workshop, which had been blown a distance of eight miles before imbedding itself in a crevasse. Items of value were salvaged, and the party spent two hours searching badly crevassed terrain for the missing aircraft.

On the 11th, it was possible for a party to return to Mawson. The tractor transported Kitchenside, Assender and Murphy part of the way, but they had to complete the remainder of the trek on foot. For those who remained at Rumdoodle, there was ample evidence of the power of the blizzard:

The remains of the Beaver, laying on its side on the airfield, looked pitiful. It is just a broken and twisted fuselage, wings are missing and gear scattered everywhere. Two concrete blocks weighing 2

> hundredweight each, with which we tied down the wings, are laying nearby, nearly 200 yards from their original position. JATO bottles in cases . . . weighing 176 pounds are a quarter of a mile from where we had them stored. Drums of fuel and oil are scattered over the plateau. The tie-down cables from the Dak. two 15 ton breaking strain and two 7 1/2 . . . give grim evidence of the holocaust. They have snapped like cotton. [97]

Remarkably, the wind fence survived.

Shortly after the trio arrived at Mawson, Mr Currie, who had been riding a motorbike on the sea ice, reported that he had found the Dakota. A dog team was harnessed, the men proceeding eight miles west of Mawson where the aircraft was lying on the heavily crevassed ice cliff. The aeroplane was:

> Approx. 400 feet above sea level. The broken ice and crevasses were covered with snow in one area that permitted an approach from the sea ice. A steep rise was ahead of the aircraft and Mr Collins (engineer) who inspected the area, said that winching out and salvage from the plateau was not possible. It would suggest that A65-81 remained intact until hitting the crevassed area ahead of the ice cliff. The starboard undercarriage collapsed first, causing the aircraft to hit the starboard wing tip and travelling this way for a short distance. The port undercarriage then collapsed and the aircraft came to rest in the crevassed ice cliff. [98]

Those who had remained at Rumdoodle returned to Mawson. On the 15th, a 15 man team, supported by a dog team and the motor bike, commenced salvage operations. A camp was established at Ringoya, on a rocky outcrop three miles from the wreck. To facilitate the salvage of equipment, a pulley block was attached to the aircraft and a line, with a sled attached, passed to the bottom of the cliff. The weather remained mild, and a full day's work saw the salvageable items stored at Ringoya recovered. Next day, the dog sled was utilized to transport the items to Mawson, to where the men, leg weary and exhausted, found their way.

The impact of the loss of the aircraft related to the safety of expedition members became alarmingly evident. News was received from the southern field party that they may have to abandon their D4 tractor, due to fuel shortages. This could place the team members in some jeopardy. The hopes of the party were twofold — that they would discover a known fuel dump, and that a relief party could meet them on their trip northward.

The object of the relief party was to proceed 150 miles inland, using the D4 tractor towing sleds complemented by a dog team, and meet the outcoming team. As the relief ship was due at the end of January, a 30 day limit was placed on the journey. If the southern party could not be sustained in that time, aerial assistance would be sought from Mirny. Felton, Machin, Kirby and Graham Dyke (as the dog team driver), volunteered for this potentially dangerous undertaking.

On 27 December 1960, the D4 tractor headed out on the first stage of the mission. Crossing meltwater streams en route, the tractor arrived at Rumdoodle, where it was joined by the dog sled party at 9.30am. A fuel dump was established within nine miles of Mawson for possible use by Russian aircraft.

Loaded with 14 drums of diesel fuel and 30 gallons of oil, the tractor team led off, heading for the gap between Mount Hordern and the David Ranges. Bad visibility caused the party to retrace its steps to Rumdoodle, where it was blizzard bound for two days. While awaiting the possible arrival of Russian aircraft, New Year's Day was celebrated with a mixture of rum and diluted grapefruit juice.

During the scheduled radio link with Mawson on the following morning the news was passed to the relief party that the authorities in Melbourne had vetoed Russian assistance, except for life-saving reasons. At 1pm the party was again crawling at 4mph through the crevasses, heading south. At 8pm 'Bech' tents were erected, and the party slept.

The recovery party. A65–81 can just be seen centre top of photograph. (Graham Dyke)

After breaking camp on the 2nd, the tractor zig-zagged through the crevassed terrain. Once it had to pull itself out of impediments to its progress, but the party reached Mount Twintop, 49 miles south of Mawson, at 3.15pm. News of the southern party was sought after camp had been established and a meal prepared. That party was 90 miles further south, but were within ten miles of a two drum fuel depot. A later message was not so heartening — the weazel had broken down and white out conditions prevailed.

Snow fell on the 3rd, and it was not possible for the relief party to travel south until the 5th. When in camp for the night, the party was advised that the southern party had repaired the weazel, discovered the fuel, and could now keep the D4 mobile. The relief party headed north on the 6th, the dogs with great enthusiasm. (99)

All that remained was for Flight members to prepare salvaged stores for return to Australia, and assist with the clearance of the station site. The thawing of the ice and snow revealed timber, drums and other debris. As part of the preparation for changeover operations, a floating jetty was constructed from empty 44 gallon drums. A close watch was kept on the movements of the relief ships *Magga Dan* and *Thala Dan.* The latter berthed, amid cheers of welcome, on 27 January 1961. *Magga Dan* joined her sister ship before sailing for Australia on 11 February; it had been the first time that both ships had been in Horse Shoe Bay at the same time.

The demise of 'Ann Cherie'. A65-81 in crevasses on the ice cliff near Mawson. (Graham Dyke)

On 13 February *Thala Dan* was pitching and rolling westward, undertaking scientific investigations along the coast of Enderby Land. It returned to Mawson on 1 March, from where it set sail for Heard Island. The ship arrived at the island on the 6th, before voyaging to Kerguelen, where it remained until the 10th. *Thala Dan* berthed at Melbourne, to the greetings of family and friends, on 22 March 1961.

The Antarctic Flight's final full year in Antarctica was over.

CHAPTER SEVEN

Summer tourists

THE loss of the aircraft at Rumdoodle had repercussions for the members who had been selected to replace the existing Flight during 1961. They were posted to other duties, and the 1961 Flight officially disbanded on 20 December 1960. The absence of any aerial presence for the 1961 season also brought adverse press comment, when the safety and viability of the expedition and its members were considered. [100] The Air Force attitude toward further deployments south was also reviewed.

The Air Board took an ambivalent view of the Air Force's commitment. It considered that little advantage would accrue to the service, as data on cold climate operations could be gleaned from Canada. Little direct benefit was gained by equipment staffs, but it was accepted that the commitment added to the service's versatility and operational techniques, as well as being good publicity. The Board accepted that:

> . . . the present situation should be accepted if the RAAF was in danger of losing prestige through handing over to another organization, but at the same time we retain the belief that we are paying too much now for the benefits we receive. [101]

On 28 July, 1961 the Air Board was advised that a request had been received for the provision of a crew and servicing team to accompany the 1962 relief expedition. [102] An affirmative response was given, and the Flight was officially raised on 2 October 1961.

The Flight consisted of N.F. Ashworth, G.G. Cooper, A.K. Richardson and R.J. Frecker. The aircraft with which it was supplied was Beaver A95-205 (ex VH-PGL) which was received from De Havillands at Bankstown on 7 October. The aircraft was used for floatplane training, and it was while on this task that Ashworth damaged a float strut when landing on rough water. The damage was repaired, and he and Richardson departed for the Gippsland Lakes near Paynesville on 2 December, to complete training. Training was initially hampered, due to the fact that:

> (in) the lateness in the decision to form the Flight many organizational details were not fully resolved beforehand. As a consequence too much time had to be spent on organization, and too little left for a comprehensive training programme . . . (103)

Members of the 1962 Antarctic Flight. L-r: Frecker, Richardson, Cooper, Ashworth. (Garry Cooper)

The Flight was considered proficient when it boarded the *Thala Dan* and sailed from Melbourne on 22 December 1961. *Thala Dan* anchored off Lewis Islet after an uneventful voyage on 29 December. Ashworth and Cooper flew photographic sorties from this base, along the coast east of the islet to Rock X, and to the Dibble Glacier tongue. On the 30th, the Beaver flew 120 miles to the west, to Cape Mose. After being grounded for two days due to a blizzard, the Beaver flew a radar heighting sortie which took it 80 miles inland — and also meant that it landed at midnight, with the sun just over the horizon. En route to Cape Carr on the following day, *Thala Dan* was the base for another Beaver flight, this time to photograph Cape Goodenough and the May Glacier tongue.

The ship moored at Wilkes on 10 January 1962, and the floatplane was secured adjacent to the camp area. From this site, the Beaver flew a reconnaissance of the Vanderford Glacier, to the west of Wilkes. This flight was followed by a geological flight to the Ballany Islands by Ashworth.

A95-205 is stowed for the voyage south. (RAAF Official)

Cooper's later attempt to undertake radar heighting duties south of Ivanoff Head was aborted due to bad weather. No further flights were made until a spark plug change had been carried out on the 12th, with the aircraft on the water and under extremely cold conditions. On the 14th, the radar heighting run penetrated 80 miles south of Ivanoff Head. This was followed by an ice reconnaissance of the pack ice to the north and west of the base. The last flight from Wilkes was to the Totten Glacier, prior to cutting 120 miles inland from the station.

Thala Dan sailed to Chick Island, arriving on 22 January. It was intended that the automatic weather station would be serviced. While this task was being undertaken, the Beaver was flown on two three-and-a-half hour flights, one to the west of the island, and the other to a point 80 miles inland. During the following day, it flew to Totten Glacier and penetrated south of Cape Milhoylov to a depth of 80 miles.

On the 30th, the expedition came face to face with history. A party landed at Commonwealth Bay, where Sir Douglas Mawson had established his base in 1911. Two days later, the ship was berthed at Durmont d'Urville. On 2 February, the ship was located in Disappointment Bay, just east of the south magnetic pole. The Beaver was not flown until the 17th, when a photographic run was made from Penguin Point to Cape Freshfield, flying within 30 miles of the pole. The return flight was hampered by snowstorms.

After visiting Thala Island in the New Zealand Antarctic Territory on the 11th, *Thala Dan* set course for Davis, where it arrived on the 21st. In the interim, the Beaver had flown a photographic sortie from Thala Island to Williamson Head. After the ship anchored at Davis, the aircraft flew inland heighting runs at a distance of 50 miles from Williamson Head. The final flight from Davis was one of four hours, from Thala Island to Renwick Bay.

The period of transition. The Beaver shelters under the stern of Thala Dan. *Note the civilian aircraft Bell 47G on the stern of the ship. (RAAF Museum)*

Macquarie Island was the next port of call for *Thala Dan*, where it arrived on 2 March. The expedition berthed at Melbourne on 8 March 1962. For the Antarctic Flight it had been a remarkably incident free voyage, marred only by the vertical camera freezing — a problem met by the subsequent Flight, and overcome by placing a cover over the lens.[104] During the period 22 December 1961 to 8 March 1962, the Flight had undertaken 24 sorties totalling 79 hours 15 minutes flying and covered some 7,000 miles, of which 1,800 were on photographic tasks.[105]

The final operations of the Antarctic Flight were carried out during the relief expedition of 1962-63. J.A. Batchelor, G.G. Cooper, A.K. Richardson and D.D. Tiller embarked on the *Thala Dan*, after training at Point Cook. Training was marred by having to have the K17 camera mount modified to enable it to be made suitable for plotting, and by the heavy seas which made it impossible to undertake floatplane training on the bay near Point Cook. Because of the sea state, Cooper was forced to land the Beaver on the Werribee River on 28 November 1962. The river was not wide enough to enable the aircraft to be turned around, so it had to be towed out tail first next day. (106)

Members of the 1963 Antarctic Flight. L-r: Tiller, Cooper, Bachelor and Richardson. (RAAF Official)

Thala Dan departed from Melbourne on 22 December, 1962 and arrived off Dumont d'Urville on the 31st, where 'the French (were) very hospitable and provided excellent French food and unlimited quantities of wine.' (107) On 6 January the ship arrived at Lewis Islet, to enable technicians to undertake maintenance on the automatic weather station.

The Beaver flew four days later, making two flights to the Rock X area. Two Bell 47G Helicopters were employed transporting spares to the island. On the 6th, a photographic run was made by the Beaver from Cape Carr to Lewis Islet, before *Thala Dan* sailed for Chick Island.

Pack ice prevented any progress toward Chick Island two days later, and the ship set course for Wilkes. The Beaver was launched on 11 January and successfully dropped mail, beer and tractor parts to the Vostok party, which had undertaken an overland trek of 1,800 miles from the Wilkes to Vostok base and return. Garry Cooper recalls that there were high winds which resulted in the stores dropping almost horizontally. To keep the Beaver on station, he flew a series of 'S' turns, thus keeping the aircraft virtually stationary above the drop zone. (108)

The Beaver was tethered to a buoy in the shelter of Western Bay, to free the ship's hold during unloading operations. On the 13th, during the second flight of the day, a rock outcrop was discovered near Cape Waldron. The presence of such a solid object was of some significance in the changing landscape, as it could be used for astrofixes. The weather prevented more flying until the 19th, when low cloud and high winds lifted and abated, giving Cooper the opportunity to fly two glaciologists (Alistair Battye and Robin Simon) on a low level reconnaissance of the Vanderford and John Quincey Glaciers.

At 11.00am, the aircraft was flying 500 feet above the Vanderford Glacier 22 nautical miles south of Hatch Island in heavy overcast. The cloud and snow, coupled with the presence of ice crystals in the air, resulted in a loss of power in the engine. Cooper applied carburettor heat, thus gaining a temporary restoration of power, but the engine began coughing and backfiring again. Juggling mixture and carburettor heat controls, Cooper turned north for the sea, slowly losing height and power. Ten miles from the coast, a forced landing on the ice was contemplated. The ice gradient below steepened, enabling the airspeed to be increased. Half a mile from the coast, the engine cut out completely. Luckily, the glide angle and the gradient of the glacier coincided, so no attempt was made to restart the engine. A 'Mayday' call was transmitted to *Thala Dan*; Cooper handed his helmet to one of the passengers, tightened his seatbelts, and prepared for the landing.

John Batchelor, acting as flight controller in the ship's radio room, requested that the helicopters operating in the area return to the *Thala Dan* and refuel. The ship departed from Wilkes at 11.50am, and the helicopters flew off at 12.31pm.

Cooper had successfully landed in relatively ice free water off Ivanoff Head. The fire warning light was flickering on landing, so a quick check was made for fire before the aircraft was paddled toward to the shore. Half an hour later, the pilot's immersion suit proved its worth, when the aircraft had to be pushed off rocks into deeper water. The aircraft was moored to rocks in a sheltered cove, a survival camp set up and one of the passengers posted at the summit of Ivanoff Head to signal rescuers.

Radio contact was not made with the helicopters until 12.50pm, due to the high terrain surrounding the Beaver. After landing at the top of Ivanoff Head, the helicopters flew the doctor and one passenger back to the ship, the other remaining with Cooper to assist in aircraft handling. Alan Richardson flew in on the return flight to check the Beaver's engine. The increasingly rough sea state resulted in the aircraft being washed against rocks, so the decision was made to taxi it into open water. The spluttering Beaver was taxied six miles to the ship, and hoisted aboard at 4.30pm. (109) A new carburettor was fitted, and the aircraft was ready for an ice reconnaissance flight on 2 February.

On 30 January, another attempt was made to reach Chick Island. The ship had been ice bound off Wilkes and the heavy pack ice barred her passage to the island, and also curtailed an approach to Porpoise Island. The ship returned to Wilkes.

The Beaver shelters near Ivanoff Head after force landing. (RAAF Official)

Thala Dan finally arrived at Porpoise Island on 4 February. The Beaver flew a photographic sortie on the 5th, which was marred by ice getting inside the camera port glass. The Beaver flew from Chick Island to Wilkes on the 8th, to acquire badly needed spares for the weather station. White out conditions over Cape Waldron forced the aircraft to return to Wilkes, and the crew stayed overnight before returning to Chick Island at first light. The Beaver was then employed on a four hour photographic flight over the area.

The ship sailed for the Mertz Glacier, but was held fast for 12 days in the pack ice encountered en route. After freeing itself from the ice on 22 February, *Thala Dan* set course for Oates Land. The Beaver was not flown until the 27th, when the pitot tube was damaged when the aircraft landed. This was repaired but a second flight was aborted when the aircraft suffered a radio failure. The sea spray had frozen to the tailplane, under surface of the wings and the floats during take-off — the conditions were totally unsuitable for floatplane operations.

Thala Dan was hampered by a blizzard while en route to Macquarie Island where it arrived on the 28th. Here it was intended to use the Beaver to photograph the island. The sea state was such that the take-off run was aborted. The eight foot swell made the recovery of the aircraft arduous. This was the final attempt to fly the Beaver in southern waters. *Thala Dan* arrived back in Melbourne on 11 March 1963, and 'all members of the RAAF Antarctic Flight proceeded on leave and on 5th April, the Flight was officially disbanded.' [110]

The final two outings for the Antarctic Flight were symptomatic of the attitude of the Air Force toward the task. It was evident that the controlling authority was of the opinion that the deployment of the Flight was an action which did not fit with their conceptions of the role of the Air Force. It must be conceded that the Vietnam commitment was growing, as was the Australian presence in Malaysia. It is significant that, during 1962, the Flight shared the stage with civilian helicopters. The use of civilian helicopters and aircraft, flown by pilots who had gained their initial Antarctic experience with the RAAF, was the pattern which has been followed ever since. Without denigrating the organizations or the men concerned, none of these deployments have had the significance or impact as did the Antarctic Flight. That flight was a credit to the RAAF, and to Australia. The air defence of Australia is the prime role of the RAAF, but it would appear just as valid to deploy a minor unit in support of Australian policy in a field which may not appear to be related to the defence of the nation. There are precedents for such a role in the short history of the RAAF — the RAAF experience in the Antarctic is one of them.

Prior to the official disbandment of the Antarctic Flight, some consideration was made to further RAAF activity over the frozen south. During late 1960 Wing Commander W. Addison, the staff officer for transport at Headquarters Operational Command, served with the American forces as an observer during Operation 'Deep Freeze'. He visited the south pole twice, the Byrd and Wilkes bases, as well as the Beardmore Glacier. On his return, he broached the subject of utilizing U.S. Air Force Hercules transports to extend the scope of Australian research and exploration in the Antarctic. [111]

The option of using American Hercules aircraft was not taken up, but two RAAF squadrons have continued the tradition of support to the Australian presence in the Antarctic and sub Antarctic islands. No.11 Squadron Orion maritime reconnaissance aircraft have flown over the Antarctic on several occasions. On 3 December, 1969 A3-295, captained by Wing Commander R.N. Law, with Air Commodore G.H. Steege as his co-pilot, flew a high altitude navigation exercise to Casey — the renamed

Wilkes site. With seven navigators and eight members of the media, the aircraft flew within 70 nautical miles of the base on a round trip of 4,200 miles. (112) A flight to Macquarie Island was made by the same squadron on 21 March 1970, when Flight Lieutenant L. Fisher and his crew dropped mail to expedition members. The 2,900 mile round trip took nine hours to complete, and the storepedo was dropped from an altitude of 300 feet to the waving inhabitants of the island.

*Orian overflys the scientific station at Macquarie Island, 7 September 1977. (*RAAF News*)*

Another visit was made on 7 September, 1977. This flight, captained by Flight Lieutenant R.D.M. Lovell, was made with the object of dropping storepedoes of meat, fresh vegetables and perishables to the Antarctic Division staff on the island. The aircraft deployed to RAAF Base East Sale for the flight. Senator Michael Townley and the Officer Commanding RAAF Base Edinburgh, Air Commodore B.J. Connaughton, with seven members of the media, returned to Edinburgh after a round trip of eight hours. The positive effect of the flight on the morale of the members of the Antarctic Division at Macquarie Island was evident, for:

> You could tell the Macquarie (radio) operator, Miss Sarah Stephens . . . was becoming more excited as the flight approached . . . (she said that) 'it's the only thing interesting we have had happen to us for quite a while.' (113)

Australian Hercules aircraft and crews have also been utilized in Operation 'Deep Freeze', the American aerial resupply of their southern scientific bases. The first Australian aircraft landed at McMurdo Sound on 1 December, 1978 after an eight hour flight from Christchurch, New Zealand. The aircraft landed in a light snow storm, and was flown by Squadron Leader Stuart Dalgleish. It carried 12 passengers and 11,000 kilograms of freight. The pilot claimed that the flight was close to routine, although he did concede that radio communication, weather and the actual landing were 'all a bit hairy at times.' (114)

Hercules transport at McMurdo Sound, 1 December 1978. (Dept. of Defence Public Relations)

The support of 'Deep Freeze' consisted of four Hercules flights between Christchurch and McMurdo Sound, in conjunction with aircraft of the U.S. Air Force, Navy and RNZAF. To enable maximum utilization, four crews were deployed. Operations had to be undertaken in November, before the ice runways commenced decaying in late December and early January. Training flights were undertaken, and Antarctic survival courses attended before the operations were flown.

Antony Underwood, a Defence Public Relations Officer, filed the story of one of these flights. With 12 members of the Antarctic survival course, he was delayed at Christchurch for three days, waiting for clear weather at the terminus of the flight, Ross Island. On 24 October 1980, Squadron Leader Bill Mattes flew over low cloud to the island. Some hours flying time from touch down, the cabin temperature in the aircraft was lowered, and the passengers donned balaclavas, gloves, woollen socks, and mukluk snow boots in preparation for the Antarctic weather. Underwood wrote that the aircraft was found to have suffered:

All the problems we get in operating in very cold weather — icing and freezing of components such as the engine bleed valves which supply warm air for the air conditioning and pressurization system. We worked pretty solidly for the four hours we were on the ice, particularly the two flight engineers — W.Off. Barry Dean and F.Sgt Marty Klein-Essink. Barry went close to suffering from frostbite because the bleed valves had to be manually operated from the ground. This involved getting the engine cowling open which was a very difficult task as the countersunk bolts had frozen into the metal. The auxiliary power unit, which has an oil reservoir fairly close to the aircraft skin, refused to start. Apart from this, a hydraulic line burst on the way down and had to be replaced at McMurdo, but that was not really a cold weather problem. (115)

Hercules crew, l-r: Sgt De Boer, Flt Off. Beaman, Sqn Ldr Dalgleish, Plt Off. Morris and W.Off. Heffernan. (Dept. of Defence Public Relations)

The Orion and 'Deep Freeze' Hercules flights attracted publicity, whereas flights by the Hercules transports of No.36 Squadron to Macquarie Island have gone unannounced. Flight Lieutenant I.G. Hawke flew A97-001 on the initial flight on 26 March, 1979 successfully dropping nine compacts of stores to the inhabitants of the island. At the time of writing (September 1989) the squadron has made 28 flights over the island; the last recorded by Flight Lieutenant A.E. Kempnich and his crew on 29 June, 1989. The flight of 14 August, 1986 is an example of the links between the Air Force and the Antarctic Division. One of the four passengers aboard the aircraft was Ian Jacobsen, who was to be the officer-in command of the Macquarie Station during the 1987 season — an ex-RAAF Wing Commander and Commanding Officer of No.34 Squadron. No.36 Squadron has undertaken this task with dedication, in all weathers, without any publicity or drama.

The RAAF is upholding the tradition established by Douglas and Campbell half a century ago. The Antarctic Flight has overflown most of the coastline of the Australian Antarctic Territory, and its role with the various expeditions cannot be overrated. They were pioneers in the field, and it appears that no one has yet surpassed their aerial contribution to our knowledge of the great white continent to our south. In following that tradition, the Air Force is also proving that it is capable of reacting to a broad definition of the role of air power in our national and international policy. Defence of national integrity by military means is an undeniable role of the military forces, but it must be argued that national aspirations can be enhanced by the peaceful use of that military asset. The actions of the Antarctic Flight are a fine example — one which may, some time in the future, be seen as a precedent for the peaceful use of airpower in the development of the Australian Antarctic Territory.

APPENDIX ONE

Training the Antarctic Flight

It became the custom, initiated when Peter Clemence travelled south with the 1956 expedition, that the commanding officer of the subsequent Antarctic Flight travelled south with the current expedition. This procedure gave the officer first hand experience in facing some of the problems which could arise during the following year.

The Flight usually commenced training during August. In addition to becoming familiar with the aircraft and other equipment, the Flight members undertook a week's survival training, which was usually undertaken at the RAAF Base Laverton ski lodge, 'Yeti', at Mount Bulla in the Victorian Highlands. The Flight members were taught to ski.

Photographic training was undertaken at the School of Photography, RAAF Base East Sale, Victoria. Here the members learned the technique of photographing from the Beaver, camera installation, film loading and processing, the use of drift sights and the plotting of vertical photography. While at East Sale, training was also given in polar navigation.

Another aspect of the training was a week of lectures at the School of Aviation Medicine at Point Cook, covering health problems to which the Flight could be subjected.

APPENDIX TWO

Antarctic airfield

Some consideration was given to the establishment of an airfield in the Antarctic, capable of being used for operations throughout the year. On 27 March, 1959 H.O. Wilson completed a report on such a site. Full communication and meteorological facilities would be required.

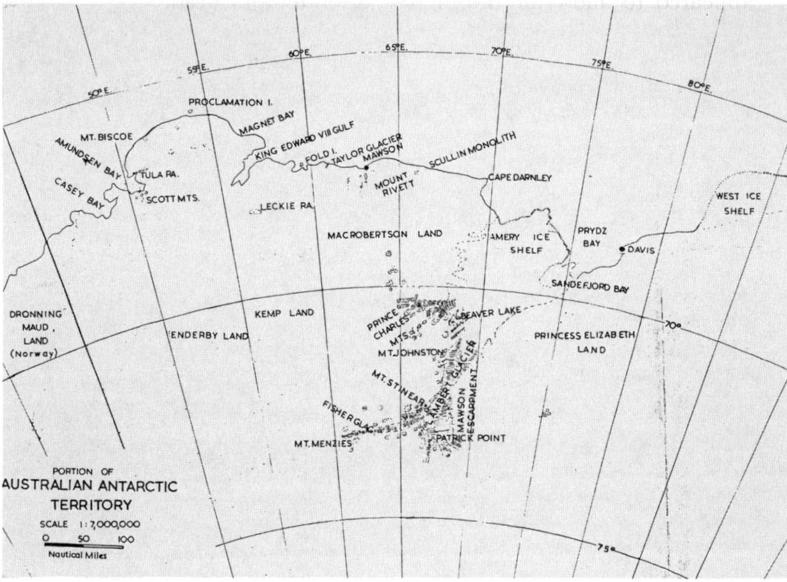

The area surveyed was adjacent to the base at Davis. The climate at the site was relatively mild by Antarctic standards, and it was roughly central between Mawson (341 nautical miles to the west), and Wilkes (741 nautical miles east). The Russian base at Mirny lay 360 nautical miles to the east. There was also the advantage that the Vestfold Hills were not subject to snow cover, thus decreasing the incidence of white out conditions on the approach and take-off from the proposed strips.

Wilson identified three options. There was the Local Strip, approximately 1½ miles north-east of the station, which was orientated into the prevailing wind. The maximum length available was 900 yards, but a rock projection would have to be removed. The second site was at the beach, 1½ miles north-east, which could be lengthened to 680 yards, in a glacial valley. The third site was also in the glacial valley, stretching from the beach to Lake Dingle. This gave 1½ miles of available runway. A land airfield could be constructed, which would prevent ablation problems, and be capable of operating aircraft to the size of a Super Constellation. Such an aircraft would have the necessary range to return to Australia without refuelling.

All three sites were readily accessible by DUKWs, an important factor when all stores, fuel and equipment would have to be supplied by sea.

Sites on the Wilkes and Mawson plateaus were considered, but the Davis site appeared to allow for the highest aircraft utilization.

APPENDIX THREE

Antarctic flight personnel 1948-1963

1948

Squadron Leader	R.H.S. Gray
Flight Lieutenant	M.D. Smith
Warrant Officer	G.C. Dunlop
Warrant Officer	P. Swan
Petty Officer	T.W. Lidell (RAN)
Corporal	R.D. Jones
Leading Aircraftman	B. Meek
Leading Aircraftman	C.E. Short

1954

Flight Lieutenant	D.W. Leckie
Flight Sergeant	R. Seaver
Sergeant	F. Morgan
Sergeant	K.W. Duffell

1956

Squadron Leader	D.W. Leckie
Flying Officer	J. Seaton
Sergeant	G. Johansen
Sergeant	G. Sundberg

1957

Flight Lieutenant	P. Clemence
Flight Lieutenant	D. Johnston
Flight Sergeant	R. Pickering
Corporal	N.W. Merideth

1958

Squadron Leader	I. Grove
Flight Lieutenant	H.O. Wilson
Sergeant	G.K. Downer
Sergeant	S.A. Manning
Sergeant	A.K. Richardson
Sergeant	O. Maguire

1959

Squadron Leader	J.C. Sandercock
Flying Officer	G.A. Banfield
Sergeant	R. Rippon
Sergeant	H. McIntyre
Sergeant	S. Bell
Squadron Leader	D.W. Leckie
Corporal	N.W. Meridith
Corporal	J.S. Williams

1960

Squadron Leader	J.C. Kitchenside
Flight Lieutenant	E.C. Bloomfield
Flying Officer	N.W. Hanson
Flying Officer	G. Dyke
Pilot Officer	K. Assender
Flight Sergeant	H.V. Carne
Sergeant	K. Felton
Sergeant	D.V. Monks
Sergeant	R.D. Murphy
Corporal	H.B. Harris
Corporal	J.T. Arthur
Leading Aircraftman	B.P. Rutter

1961*

Squadron Leader	J.R. Batchelor
Flying Officer	B.J. Lofthouse
Flying Officer	R.M. Raynes
Pilot Officer	D.D.R. Marks
Flight Sergeant	R.L. Marschke
Sergeant	R.B. Hope
Sergeant	P.B. Kemp
Sergeant	L.R. Lowe
Sergeant	V. Varney

1962

Squadron Leader	N.F. Ashworth
Pilot Officer	G.G. Cooper
Sergeant	A.K. Richardson
Corporal	J. Frecker

1963

Squadron Leader	J.R. Batchelor
Pilot Officer	G.G. Cooper
Flight Sergeant	A.K. Richardson
Sergeant	D.D. Tiller

*Due to the loss of the aircraft during the 1960 expedition, those listed did not travel to the Antarctic during 1961.

APPENDIX FOUR

Antarctic flight aircraft

SIKORSKY KINGFISHER

A48-13 photographed at Adelaide during 1947. It is possible that trials were undertaken to check the feasibility of carrying the aircraft aboard Wyatt Earp *which was based at Port Adelaide at this time.*

POWERPLANT
1 450 h.p. Pratt and Whitney Wasp junior R-985-AN2 Radial
DIMENSIONS
Span, 35ft $10\,^{11}/_{16}$*in (10.96m), Length, 33ft* $7\,^{3}/_{16}$in (10.25m), Height, 14ft 8in (4.47m), Wing Area 261.9sq ft (24.4sq m)

WEIGHTS
Empty, 3335lb (1514Kg), Loaded, 4980lb (2260.8Kg)
PERFORMANCE
Maximum speed at 5000ft (1525m), 171 m.p.h (275 K/hr), Cruising speed at 6000 ft (1830m) at 75% power 152 m.p.h. (244.7 K/hr), Rate of climb at 4000ft (1220m) 960ft/min (292.8 m/min), Service ceiling 18200ft (5550m), Cruising range at 6000 ft (1830m) at 75% power, 908 miles (1460 Km)

Only one Kingfisher operated in the Antarctic, A48-13. This aircraft was delivered to the RAAF on 6 May 1942, and was on the strength of No.107 Squadron. Placed in storage after the war, the aircraft was refurbished and issued to the Antarctic Flight on 28 November 1947. When it returned from the Antarctic, the aeroplane was stored at Rathmines, New South Wales, for possible further issue to ANARE. In the event, it was not required again, and the airframe was declared for disposal on 21 March 1953.

VICKERS SUPERMARINE WALRUS

HD 874 tethered at Heard Island. LST 3501 is unloading in the background. (RAAF Museum)

POWERPLANT
1 750 h.p Bristol Pegasus Radial
DIMENSIONS
Span, 45ft 10in (13.97m), Length, 37ft 7ins (11.45m),Height, 15ft 3in (4.65m), Wing Area, 610Sq ft (56.67Sq m).

WEIGHTS
Empty, 4900lb (2223Kg), Loaded, 7200lb (3266Kg).
PERFORMANCE
Maximum speed at sea level, 124 m.p.h. (200 Km/hr),Cruising speed at 3500ft (1070 m), 95 m.p.h. (153 Km/hr), Initial climb rate, 1050ft/ min (320 M/min), Service ceiling, 18500ft (5640m), Cruising range (94 m.p.h at 3000ft) 600 mls (966Km)

The only Walrus to serve with the Antarctic Flight was HD874, which was delivered to the RAAF on 14 September 1943. The aircraft served with No.9 Squadron and No.8 Communications Unit, before being placed in storage. After refurbishment at Lake Boga, Victoria, the aircraft was issued to the Antarctic Flight on 3 November 1947. Lost in a gale at Heard Island during December 1947, the aircraft was recovered and is now undergoing restoration at the RAAF Museum Point Cook.

AUSTER MK 6

A11-200 on the ice during the approach to Mawson, January 1955. Note the Weazel to the right of the aircraft. (Author's collection)

POWERPLANT
1 145 h.p De Havilland Gipsy Major 7
DIMENSIONS
Span, 36ft (11.0m),Length, 23ft 9in (7.23m) Height, 8ft 4 1/2in (2.55m), Wing area, 184Sq ft (17.1Sq m)

WEIGHTS
Empty, 1469lb (666.2Kg), Loaded, 2122 lb (962Kg)
PERFORMANCE
Maximum speed, 122 m.p.h (191.2 Km/hr), Cruising speed, 107 m.p.h (171.2 Km/hr), Initial climb rate, 660ft/min (201M/min) Service ceiling, 12000ft (3660m), Range 315 mls (504 Km).

Two Auster Mark 6 aircraft (VX-126 and VX-127) operated with the Norwegian-British-Swedish Antarctic Expedition of 1949-52. These aircraft were purchased for use by ANARE, and became A11-200 and A11-201 respectively. They were held in storage until the 1954-55 expedition. The aircraft suffered damage and were re-built as a single aeroplane, known as A11-200. This aircraft was lost overboard after fouling a lifeboat on the *Kista Dan*, and the remains of A11-201 were returned to No.1 Aircraft Depot, Laverton. The aircraft was then transported to the Royal Victorian Aero Club on 19 April 1954. The Antarctic Division of the Department of External Affairs re-purchased the aircraft, which was re-issued with the serial number A11-201. The Auster was overhauled by Kingsford Smith Aviation at Bankstown, NSW prior to being fitted with a Gipsy Major Series 10 Mk 2 engine at De Havillands. The aircraft saw service with the Antarctic Flight during 1956-57 and with the Wilkes relief expedition of 1959. The aircraft was then disposed of, and became VH-RCT. It operated out of Flinders Island before being written-off in a landing at Cape Barren Island on 5 February 1964.

De HAVILLAND DHC-2 BEAVER

A95-202 shows her ski undercarriage and photographic equipment. (Dept. of Defence Public Relations)

POWERPLANT
1 450 hp Pratt and Whitney R-985 Wasp Junior
DIMENSIONS
Span, 48ft (14.64m), Length, 30ft 3in (9.22m) Height, 9ft (2.75m)

WEIGHTS
Empty, 2960lb (1343Kg), Loaded, 5100lb (2300Kg).
PERFORMANCE
Maximum Speed at sea level, 163 m.p.h (262 Km/hr), Cruising speed at 5000ft (1525m) 140 m.p.h (225 km/hr), Initial climb rate, 1020ft/min (311 M/min), Service ceiling, 19000ft (5797m), Maximum range, 740 mls (1190 Km)

In total, four Beavers were operated with the Antarctic Flight, serial numbers A95-201 to A95-203 and A95-205. A95-201, A95-202 and A95-203 were lost to the elements. A95-201 and A95-203 were recovered and re-built as one aircraft, which was registered as VH-AAV. Prior to being taken over by the Antarctic Flight in 1960, A95-202 had been registered as VH-PGL. This aircraft was written-off during the December blizzard of 1960. The second VH-PGL became A95-205, and later served for six years with the Snowy Mountains Hydro Electricity Authority, as VH-SMD. The aircraft was then purchased by Aerial Agriculture Pty Ltd and was later sold to the United States as N541R on 3 June 1968. A95-204 was operated on loan from De Havillands by No.1 Air Trials Unit from 15 December 1959 until 3 August 1961.

DOUGLAS DAKOTA

A65-81 undergoing JATO trials at ARDU. Note the skis and the camera mounts to the rear of the fuselage. (RAAF Museum)

POWERPLANT
2 1200 h.p. Pratt and Whitney R-1830-S1C3G Radials.
DIMENSIONS
Span, 95ft (28.9m), Length, 64ft 5 1/2in (19.63m), Height, 16ft 11 1/8in (5.2m), Wing area, 987Sq ft (91.7m)

WEIGHTS
Empty, 16865lb (7657Kg), Loaded, 25200lb (11441 Kg)
PERFORMANCE
Maximum speed, 230 m.p.h. (368Km/hr) at 8500ft (2590m), Cruising speed, 207 m.p.h (331.2 Km/hr), Initial climb rate, 1130 ft/min (345M/min), Service ceiling 23200ft (7076m) Normal range, 2125 mls (3400Km)

The only Dakota which served with the Antarctic Flight was A65-81, which was delivered to the RAAF on 11 March 1945. It was issued to No.36 Squadron fifteen days later. The aircraft served with RAAF stations Pearce and Parafield, as well as No.86 Wing. The aircraft was damaged three times by the elements; once at Boulder airfield on 21 November 1946 when a wing tip was blown against a building in a storm. The second occasion was when she was damaged by cyclonic hailstones. The last was fatal, when she failed to ride out a blizzard at Mawson on the night of 9/10 December 1960.

APPENDIX FIVE

Awards

Members of the Antarctic Flight were awarded various awards for service in the Australian Antarctic Territory. Among them was A.K. Richardson, who was awarded a B.E.M. for his actions during the 1957-58 season, and J.C. Sandercock who was honored with an O.B.E., earned during the summer hurricane of December 1959. At least two other awards were presented to Flight Members; an O.B.E. to D.W. Leckie and an A.F.C. to I.L. Grove. The respective citations for these two officers read as follows:

> . . . In 1955, Squadron Leader Leckie was appointed Officer Commanding the RAAF Flight which accompanied the Australian Antarctic Research Expedition to Mawson in 1956. In his preparations for the expedition, this officer showed outstanding zeal and determination, and successfully overcame any difficulties. His thorough preparations contributed greatly to the success of the work performed by the Antarctic Party during 1956. En route to Mawson, the RAAF Flight engaged in numerous photographic reconnaissance flights over many miles of unknown coast in Australian Antarctic Territory and from the Base itself this Flight achieved complete photographic cover of the coast of Princess Elizabeth Land, MacRobertson Land, Kemp Land and Enderby Land. A number of flights was also made into the interior of the Antarctic Continent over terrain that was virtually uncharted and most inhospitable and where the chances of rescue were remote. During these flights, many new and important geographical features were discovered.

The aircraft of the RAAF Flight flew without mishap some 600 hours of which 175 were spent in aerial photography. They travelled in all more than 50,000 nautical miles, carrying 152 passengers and 25 250 pounds of cargo. In his leadership of the RAAF Flight and his untiring work as a pilot, Squadron Leader Leckie showed great initiative, courage and determination in the face of extreme hardships and difficulties. His abilities contributed greatly to the successful completion by the Expedition of a most valuable and extensive year's work.

Ivan Grove's citation reads:

> Whilst flying a Beaver aircraft in the Antarctic in August 1958, Squadron Leader Grove was forced to turn back to his base at Mawson because of adverse weather conditions at his destination. On the return flight the aircraft developed engine trouble and an immediate landing became necessary. The terrain in the area is rugged and crevassed;

however, Grove selected a small clear area near the coast and in spite of a 50 knot wind and severe turbulence over the ice cliffs, landed the aircraft safely.

Squadron Leader Grove and his two passengers succeeded in securing the aircraft after considerable difficulty. The next day repair equipment was flown in and Squadron Leader Grove and an airman remained with the aircraft, in an attempt to repair the engine and fly the aircraft back to Mawson. Two days later Squadron Leader Grove took off alone in the Beaver and although the engine was still malfunctioning, by skilful flying he was able to reach Mawson without further mishap.

Throughout the operation Squadron Leader Grove showed skill and resource above what is reasonably expected from a young officer. He refused to be over-awed by the obvious dangers of difficult terrain and dangerous flying conditions, and resolutely met and overcame all problems.

During his Air Force career this officer has flown a total of 3365 hours of which 173 hours have been flown in the past six months. Of his total flying time he has flown 269 hours during service in the Antarctic.

REFERENCES

1. 'The Antarctic and Australian Aviation', *Aurora: ANARE Club Journal*, vol.5, no.3, 3 March 1986, p.26.
2. A.G. Price, *The winning of Australian Antarctica — Mawson's BANZARE Voyages 1929-1931*, A.R.E. Voyages 1929-1931', Angus and Robertson, Sydney, 1962. p.21.
3. Ibid, p.34.
4. H. Fletcher, *Antarctic Days with Mawson — a personal account of the British, Australian and New Zealand Antarctic Research Expedition of 1929-31*, Angus and Robertson, Sydney, 1984, p.111.
5. Ibid, p.128.
6. S.A.C. Campbell, correspondence, 7 May 1986.
7. Fletcher, op cit, p.132.
8. S.A.C. Campbell, 'With *Discovery*, 1929-30', *The Bomber: Journal of No. 1 Squadron RAAF*, vol.1, no.1, August 1931, p.10.
9. Fletcher, op cit, p.146.
10. Ibid, pp 170-171.
11. Ibid, p.194.
12. Ibid, p.196.
13. Ibid, p.205.
14. Price, op cit, p.97.
15. Fletcher, op cit, p.229.
16. Ibid, p.267.
17. Undated report by G.E. Douglas, (RAAF Historical Section).
18. Price, op cit, p.161.
19. 'The Antarctic and Australian Aviation', op cit, p.29.
20. J. Coleman-Cooke, *'Discovery II' in the Antarctic — The Story of British Research in the Southern Seas*, Odhams, London, 1963, p.178.
21. Squadron Leader G.L. Grendon, Coulthard-Clark interview, 26 April 1989, (RAAF Historical Section).
22. Report on relief flight to Mr Lincoln Ellsworth and Mr H. Hollick Kenyon per R.R.S. *Discovery II*, January 1936, (RAAF Historical Section).
23. Coleman-Cooke, op cit, p.180.
24. G.E. Douglas, Rough log of the voyage to the Bay of Whales, Ross Sea, made in R.R.S. *Discovery II* to locate Lincoln Ellsworth and H. Hollick Kenyon, January 1936, (RAAF Historical Section).
25. G.E. Douglas, 'The Search for Lincoln Ellsworth,' unpublished account. (RAAF Historical Section).
26. Rough log entry, 15 January 1936.
27. Rough log entry, 11 February 1936.
28. Rough log entry, 16 February 1936.
29. 'Pelorus', 'Antarctic Flight and Rescue – Part 2', *Aircraft*, February 1986, p.54.
30. *Aircraft*, June 1946, p.25.
31. No. 1 Air Performance Unit, Laverton: performance report on 'Special Flight Southward of Australia' by Flying Officer R.R. Shaw and members of the crew, April 1947; and Barry Pattison, 'Seaplanes in the Antarctica 1947-48', *Journal Aviation Historical Society of Australia*, vol.17, no.6, January/February 1977, p.8.
32. CRS A705, item 73/1/4068.
33. CRS A705, item 83/1/1443.
34. Ibid.
35. A. Scholes, *Fiftenn Men*, Cheshire, Melbourne, 19489, p.14.
36. S.A.C. Campbell, diary of Heard Island expedition, 28 November-18 January 1948, p.9, (ANARE Library).
37. ibid, p.19.
38. Ibid, p.27.
39. Adelaide *Advertiser*, 25 November 1947.
40. R.H.S. Gray, 'Flying in the Antarctic', *Aircraft*, 14 June 1948, p. 42.
41. CRS A705, item 153/1/1422.
42. D. Vincent, *Catalina Chronicle*, privately published, Adelaide, 9181, p.95.
43. CRS A705, item 153/1/1422.
44. Ibid.
45. Pattison, op cit, p.13.
46. D.W. Leckie, 'Report on Operations Antarctic Flight 1954-55', (RAAF

Historical Section).

47. 'RAAF in the Antarctic', *Wings*, 1 September 1955, p.11.
48. Leckie, op cit.
49. P. Law, *Antarctic Odyssey*, Heinemann, Melbourne, 1983, p.91.
50. Leckie, op cit.
51. Ibid.
52. Ibid.
53. Ibid.
54. Ibid.
55. Law, op cit, pp.149–150.
56. Leckie, op cit.
57. CRS A1969/100, item 17/2/Air.
58. Report on the 1955 Antarctic Flight, p.3, (RAAF Historical Section).
59. Ibid, p.10.
60. Ibid, p.39.
61. Leckie, 'Air operations Mawson 1956', (RAAF Historical Section).
62. Ibid; also 'Report', P.24.
63. 'Report of the 1957 Antarctic Flight', p.7, (Copy held by Peter Clemence).
64. Ibid, p.12.
65. Ibid, p.19.
66. Ibid, P.14.
67. Ibid, p.19,
68. I. Grove, correspondence, 1 April 1986.
69. Antarctic Flight Unit History (A50), Entry for 1 February 1958, (RAAF Historical Section).
70. A. Richardson, diary extract 1–2 February 1958, (author's possession).
71. Ibid, 25 April 1957.
72. Report on forced landing of A95-203 . . . on 15 August 1958 and subsequent action, p.1, (RAAF Historical Section).
73. Richardson, op cit, 17 August 1958.
74. Ibid, 18 August 1958.
75. Antarctic Flight Unit History (A50), entry for 10 December 1958.
76. CRS A705, item 182/1/513.
77. Ibid.
78. J. Bechervaise, *Blizzard and Fire — A Year at Mawson, Antarctica*, Angus and Robertson, Sydney, 1963, p.12.
79. Ibid, p.44.
80. Ibid, pp.140–141.
81. RAAF Antarctic Flight 1958–59, Spring-Summer Operations, p.7, (RAAF Historical Section).
83. CRS A1969/100, item 3/34/Air.
84. Report by the Director, Antarctic Division: Voyage of the M.V. *Magga Dan* to Antarctica, 1959, p.3, (ANARE Library).
85. Ibid, p.6.
86. Ibid, p.7.
87. Law, op cit, p.91.
88. K. Felton, personal diary, entry 1 February 1960, (Mrs S. Felton).
89. Ibid, 24 March 1960.
90. Antarctic Flight Unit History (A50), Entry for 12 May 1960.
91. Felton, op cit, 21 April 1960.
92. Ibid, 14 May 1960.
93. Ibid, 14 August 1960.
94. Ibid, 15 September 1960.
95. Ibid, 22 September 1960.
96. ibid, 9 December 1960.
97. Ibid, 11 December 1960.
98. Antarctic Flight Unit History (A50), entry for 11 December 1960.
99. Felton, op cit, 27 December 1960–6 January 1961.
100. *Sydney Morning Herald*, 17 April 1961.
101. Air Board Record of Proceedings 4/61, 10 February 1961.
102. Ibid, 62/61, 28 July 1961.
103. Antarctic Flight, commanding officer's report, November 1961, (RAAF Historical Section).
104. G.G. Cooper, interview, 23 August 1986.
105. Antarctic Flight, commanding officer's report. 22 December 1961–8 March 1962.
106. Unidentified newspapaer extract, 28 November 1962, (RAAF Historical Section).
107. Antarctic Flight Unit History (A50), entry for 31 December 1962.
108. Cooper, op cit.
109. Cooper, op cit; 'Report on forced

landing of Beaver A95-201', (RAAF Historical Section).

110. Antarctic Flight Unit History (A50), Entry for 5 April 1963.

111. Hobart *Mercury*, 21 November 1960.

112. No. 11 Squadron Unit History (A50), entry for 3 December 1969, (RAAF Historical Section).

113. *RAAF News*, September 1977.

114. *RAAF News*, January/February 1979.

115. *RAAF News*, December 1980.

116. CRS A705, item 7/1/2477.

BIBLIOGRAPHY

Secondary sources

Books

Bechervaise, J.M., **Blizzard and Fire — A Year at Mawson Antarctica,** Angus and Robertson, Sydney, 1963.

Coleman-Cook, J., **Discovery II in the Antarctic — The Story of British Research in the South Seas,** Odhams, London, 1963.

Fletcher, H., **Antarctic Days with Mawson — A Personal Account of the British, Australian and New Zealand Research Expedition of 1929-31,** Angus and Robertson, Sydney, 1984.

Jane's All the Worlds Aircraft.

Law, P.G. and Bechervaise, J.M., **ANARE — Australian Antarctic Outposts,** Oxford University Press, Melbourne, 1957.

Law, P.G., **Antarctic Odyssey,** Heinemann, Melbourne, 1963.

Mawson, P., **Mawson of the Antarctic,** Longmans, London, 1964.

Mignot, A., **The Lonely South,** Rupert Hart-Davis, London, 1958.

Parnell, N.M. and Lynch, C.A., **Australian Air Force Since 1911,** A.H. and A.W. Reed, Sydney, 1976.

Price, A.G., **The Winning of Australian Antarctica — Mawson's BANZARE Voyages 1929-31,** Angus and Robertson, Sydney, 1962.

Swan, R.A., **Australia in the Antarctic,** Melbourne University Press, Melbourne, 1961.

Scholes, A., **Fourteen Men,** Cheshire, Melbourne, 1949.

Vincent, D., **Catalina Chronicle,** privately published, 1981.

Australians in the Antarctic, Antarctic Division of the Department of External Affairs, Melbourne, un-dated.

Periodicals and newspapers

Aurora (ANARE Club Journal), 3 March 1986.

Aircraft (various dates).

Journal, Aviation Historical Society, Vol.17, No.6.

Wings (Official Journal of the RAAFA) (various dates).

Hobart Mercury.

Melbourne Age.

Brisbane Courier Mail.

Pix magazine.

RAAF News.

Rag and Tube (the newsletter of the antique aeroplane association of Australia).

The Royal Australian Air Force Reserve Magazine.

Triad (a Defence Magazine) No.12, Summer 1978.

The Bomber, The Journal of No.1 Squadron RAAF, Vol.1 No.1, August 1931.

Primary sources

Official sources

Department of Defence (Air Force Office)

Unpublished Report by Group Captain Eric Douglas (Rtd) (Incident involving Sir Douglas Mawson 1931).

Report on relief flight to Mr Lincoln Ellsworth and Mr H Hollick Kenyon per R.R.S. Discovery II, January 1936.

Rough log of the voyage to the Bay of Whales made in R.R.S. Discovery II to locate Lincoln Ellsworth and H. Hollick Kenyon after his trans-Antarctic flight January 1936.

ABC 'News Review' transcript 15 March 1947.

Report on Operations Mawson 1956.

Report on Operations Antarctic Flight 1954–55.

Report on the 1956 Antarctic Flight.

Report on the 1957 Antarctic Flight.

Antarctic Flight Unit History (A50)

1957/58, 1961/62 and 1962/63.
Antarctic Flight, Commanding Officers Reports 1961/62.
RAAF Directorate of Technical Services, No.1 Air Performance Unit, Laverton. Performance report on 'Special flight southward of Australia', by Flying Officer R.R.Shaw and members of the crew, April 1947.
Report on the forced landing of A95-203 at 67 degrees 47 minutes south, 66 degrees 25 minutes east on 15 August 1958 and subsequent action.
Report on the 1958/59 Antarctic Flight.
No.11 Squadron Unit History (A50).
No.36 Squadron Unit History (A50).
Telex report 'Forced landing of Beaver A95-205'.
Air Board Agenda Numbers 6893, 7355 and 7392.
Air Board Record of Proceedings 67/49, 102/57, 4/61 and 62/61.

Department of Science Antarctic Division

Stuart Campbell, diary 28 November 1947-8 January 1948, Heard Island Expedition.
Voyage of M.V. Magga Dan to Antarctica 1959. Report by the Director of the Antarctic Division.
Antarctic Flight Unit History (A50) 1959/60.

Australian Archives

CRS A1969/100

Base Squadron Point Cook.
3/20/Air 'F.O.M. Antarctic Flight 1956/57'.
3/27/Air 'F.O.M. Antarctic Flight 1957/58'.
3/29/Air 'F.O.M. Antarctic Flight 1958/59'.
3/34/Air 'RAAF Antarctic Flight 1959 (Wilkes)'.
3/37/Air 'F.O.M. Antarctic Flight 1959/60'.
3/41/Air 'F.O.M. Antarctic Flight 1960/61'.

RAAF Station Point Cook
3/20/Air 'Antarctic Flight Point Cook'.

Antarctic Flight
3/1/Air 'F.O.M. Antarctic Flight'.
42/1/Air 'F.A.G. Antarctic Flight'.

Maintenance Command

17/2/Air 'RAAF participation in Antarctic Expedition'.

CRS A705

7/1/2477 'Selection of Airfield — Antarctic 1959'.
73/1/4068 'Preparation and arrangements for Australian Antarctic Expedition — Issue of stores and equipment'.
83/1/1445 'Australian Antarctic Expeditions financial arrangements 1947/58'.

Private sources

Felton, K., Personal diary of 1959-60 Antarctic Flight. Richardson, A.K., Extracts from personal diary 1957-58 Antarctic Flight.